VINTAGE PLACES
A CONNOISSEUR'S GUIDE TO NORTH AMERICAN
WINERIES AND VINEYARDS

VINTAGE PLACES

A CONNOISSEUR'S GUIDE TO NORTH AMERICAN
WINERIES AND VINEYARDS

by Suzanne Goldenson

THE MAIN STREET PRESS · Pittstown, New Jersey

First edition 1985

Published by
The Main Street Press, Inc.
William Case House
Pittstown, NJ 08867

Published simultaneously in Canada by
Methuen Publications
2330 Midland Avenue
Agincourt, Ontario M1S 1P7

Printed in the United States of America

Designed by Frank Mahood.

Library of Congress Cataloging in Publication Data

Goldenson, Suzanne, 1944—
  Vintage places.

  Bibliography: p.
  1. Wine and wine making—United States. 2. Wine
and wine making—Canada. I. Title.
TP557.G65   1985   663'.22'097   85-8775
ISBN 0-915590-76-X
ISBN 0-915590-71-9 (pbk.)

Contents

Introduction

A very quiet, yet beneficial, revolution has been sweeping across the United States and Canada over the last twenty-five years—a viticultural revolution, which had its beginnings in California in the 1960s. Residents of states and provinces as far-flung and climatically diverse as Texas, Virginia, Connecticut, Washington, Missouri, and Ontario, following California's lead, reexamined their grapegrowing potentiality and began replanting their agricultural acreage and all-Labrusca vineyards to premium wine grapes. These entrepreneurs then petitioned their respective governments to pass farm winery legislation which would enable them to produce and sell wines from their new crops.

Now many of the fruits of this viticultural trend are reaching our dinner tables and are already challenging the best wines that Europe has to offer. *Vintage Places* is intended as a guide to some of the wineries that have led the way and to some that are just beginning production of their first vintages. More than 200 such wineries are included here; they are located in all parts of the United States and in three Canadian provinces, so that there are bound to be some that are accessible to every North American traveler. *Vintage Places*, however, does not pretend to be a complete guide to wineries; there are more than 600 in California alone—enough for several such volumes. The selection included, therefore, represent my choices, choices based on four general criteria: (1) the wineries included make premium (usually award-winning) wines; (2) have attractive facilities that are both hospitable and educational to visit; (3) are estate wineries (those with vineyards on the property), although there are a few exceptions; and (4) have vineyards planted to vinifera grapes—the superb European varieties such as Chardonnay and Cabernet Sauvignon.

If you have never visited a winery, a wonderful experience awaits you. Most wineries, whether owned by a large corporation or an individual, enjoy an unparalleled tradition of hospitality. You'll be given a guided tour of the working winery (and sometimes its vineyard, as well) and treated to a tasting of its products. Many of the establishments included here augment these amenities with handsomely landscaped picnic grounds, barbecue facilities, wagon rides, art exhibitions, special festivals, and other events throughout the year, including concerts, dinners, and even cooking classes.

As you would expect, the larger, better-known wineries in the most popular grapegrowing regions (such as California's Napa and Sonoma Valleys and New York's Finger Lakes) have the most extensive and elaborate visitor facilities. But these large-scale attractions are also, predictably, the most crowded. You will frequently get more personal attention at a smaller winery in one of the newer viticultural regions such as Long Island or Oregon, where the owner/winemaker may serve as your tour guide and pour the wines in the tasting room. Consequently, in each region I have tried to include a selection of wineries of all sizes and ages, both to give you a choice and to let you make comparisons among them should you be fortunate enough to visit more than one.

It may be advisable to visit the largest and most popular wineries in the off-season. For example, the Napa Valley traffic can be bumper-to-bumper during the harvest crush, but during the mild winter months, when the vineyards are at their greenest and yellow mustard is blooming amid a sea of trellised vines, you will have your pick of the region's wineries, restaurants, and inns.

There are various approaches to touring wineries, but the general consensus is that you should plan to visit no more than three or, at most, four separate ones in a single day. And if you do plan to stop at more than one, experts suggest tasting only one wine at each location, such as Chardonnays in California or Seyval Blancs in the East, so that you can compare the effects of different growing conditions and aging methods. And while many wineries do offer picnic supplies and snacks, you'd probably be wise in the outlying areas to pack a hamper and bring along a blanket so that you can linger awhile to enjoy the scenery.

It goes without saying that your visit to most wineries will give you the opportunity to purchase some of their products to bring home (the exceptions, based on local regulations, are noted in the individual entries and in the introductions to the regions in which they occur). Most state laws allow a ten-percent saving on case lots, and some wineries will allow you to mix the selection. One of the side benefits of your visit may be the opportunity to buy wines that are not generally available outside the area in which they are made, either because of local regulations or because they are too fragile to travel well.

We have taken several liberties in the design and format of *Vintage Places* in order to make it easier to use. Shorter entries appear in italics either because the wineries were lacking in one of the major criteria used for general selection but are worth a visit in any event, or because they are in an area so densely populated with wineries that space limitations demanded a shorter description. To make the entries easier to read, the names of all wine grapes are capitalized, an editorial practice that will enable you to find a particular variety more easily should you be interested in just one or two. And although many North American wineries produce what they call champagne, we have bowed to the French convention which dictates that the term be used only for sparkling wines produced in that region of France. Such wines, therefore, are called sparkling wines unless we are referring to a particular winery's appellation for its product.

It is my pleasure to acknowledge the help of some experts in the field of viticulture and vinification. Leon Adams, author of the landmark book *The Wines of America*, was most generous with his time and advice. Rory Callahagn, the New York State representative of the California Wine Institute, gave me a great deal of guidance on Empire State wineries. In addition, there were many regional winemakers and connoisseurs whose writings and perspectives were invaluable in the preparation of this book.

New England

In the past ten years, New England, never regarded as an important winegrowing region of America, has been quietly undergoing a viticultural revolution. In that one decade, ten of the region's seventeen bonded wineries have been established and hundreds of acres planted to vineyards. Where Labruscas once grew wild, premium wine grapes—viniferas and French-American hybrids once thought unable to survive the bitter New England winters—are thriving and producing prize-winning wines.

The potentiality was always there. That wild grapes grew abundantly in early New England is well documented. More than 300 years ago the Pilgrims fermented the native grape to make wine for the first Thanksgiving feast. The Concord grape, cultivated in the mid-19th century, is named for the historic Massachusetts town established some 200 years earlier. Connecticut's state seal includes a cluster of grapes in tribute to a once important agricultural crop. And the island of Martha's Vineyard was named for the wild vines that once covered its shores. Despite New England's proven track record of growing Labruscas—largely for juice and jam—the area's inhospitable climate discouraged ambitious winegrowing until the early 1970s, when plantings of premium grape varieties began taking hold.

Recent experimentation with hardier grape varieties, a more scientific understanding of climatic influences and soil conditions necessary for the nurturing of premium wine grapes, and the passage of farm winery bills are the factors contributing to the changing winegrowing picture in New England. And while there are wineries as far north as Maine and as far inland as New Hampshire, leading the way are the states of Connecticut, with seven wineries, and Massachusetts and Rhode Island, each with four.

As one would expect, most of the new wineries and vineyards are springing up along coastal areas which benefit from moderating ocean breezes. Crosswoods in Connecticut and Sakonnet in Rhode

Island are examples of this trend. (Rhode Island's Narragansett Bay has the mildest climate in the Northeast.) Taking even greater advantage of the ocean's temperate effects are those vineyards established offshore on islands, such as Prudence Island Vineyard in Rhode Island's Narragansett Bay, and Chicama Vineyards off the coast of Massachusetts. Others have taken root inland where favorable climates for grape cultivation have been identified, such as Haight in Litchfield, Connecticut, and Hopkins Hill in the Berkshire foothills.

The wines produced from New England's fledgling vineyards are not yet widely distributed nor are they well known, but among wine critics and connoisseurs they are gaining recognition (holding their own against their California counterparts), and being added to the wine lists of the finest restaurants in the Northeast. Despite these developments, premium winemaking is still in its infancy in New England. Clearly, the best is yet to come . . . and to be tasted.

CONNECTICUT

Litchfield

Haight Vineyard & Winery
Chestnut Hill
Litchfield, Connecticut 06759
Telephone: (203) 567-4045
Owner: Sherman P. Haight

Situated on a hillside overlooking the historic and picturesque town of Litchfield, Haight Vineyard & Winery is surrounded by forty-five acres planted to Chardonnay, Johannisberg Riesling, Pinot Noir, Seyval Blanc, and Maréchal Foch.

Established in 1975, Haight has a handsome new combined winery and tasting/retail facility, its entranceway enhanced by a large, specially commissioned stained-glass panel. Guided tours of this two-story structure are quite complete; they take about forty-five minutes and include a short video presentation. Also shown are the functional aspects of the working facility: its bottling line, bottles *en tierage,* and riddling racks, which are integral parts of the traditional *methode champenoise* process; and the aging cellar and fermenters. Upstairs, an attractive tasting room/retail area includes two fireplaces, cozy amenities on a chilly day. And a balcony off this second-story level offers a sweeping view of the vineyards. Outside the winery, a vineyard walk is punctuated with informative signs commenting on viticultural practices and the grape varieties grown; picnic tables are located nearby.

Haight makes 6,000 cases of wine yearly. Eighty-five percent is table wine; fifteen percent, sparkling *methode champenoise.* The product line includes the varietals: Chardonnay, Artist Series Chardonnay, Riesling, and Maréchal Foch. Generics include Covertside white and red table wines and Blanc de Blancs sparkling wine. Haight is particularly known for its Artist Series Chardonnay, Riesling, and Blanc de Blancs.

Open all year, daily except major holidays: Monday-Saturday 10:30-5:00, Sunday 12-5. Guided tours on the hour (45 minutes). Self-guided vineyard walk. Free tasting. Retail outlet: wine, gifts. Picnic facilities. Directions: From Litchfield center, one mile east on Route 118 to Chestnut Hill. Right on Chestnut Hill to winery (on your left).

New Preston

Hopkins Vineyard
Hopkins Road
New Preston, Connecticut 06777
Telephone: (203) 868-7954
Owners: William and Judy Hopkins

Hopkins Vineyard is located in western Connecticut's Warren County, between Lake Waramaug and the Berkshire foothills—an area which boasts some of the state's loveliest scenery. The charming Hopkins Inn, across the road from the winery, features gourmet cuisine and, naturally, an extensive wine list. It's a good place for lunch or dinner, or even an overnight stay.

Hopkins Vineyard is in a weathered but carefully restored 150-year-old barn on property that has been farmed by the owners' family for nearly two centuries. Old siding replaced on the exterior was recycled on interior walls, and large expanses of glass were installed to afford views of the lake. The inviting tasting room/retail area is finished with wide-board

pine flooring, a handsome chestnut tasting bar, and a sturdy cast-iron wood-burning stove. In the recesses of the barn you can see the fermenting tanks, oak casks for aging the wine, and the bottling machine. Wine awaiting sale is stored below in the cool stone cellar. Guided tours of the winery include a slide presentation in the projection room; there is also an exhibit of antique winemaking tools and equipment, and, outside, a short marked course through the vineyard itself.

Twenty acres of French-American hybrids are planted on the surrounding land. Varieties grown include Leon Millot, Maréchal Foch, Aurora, Ravat, Seyval Blanc, and Cayuga White. From these, the Hopkinses produce 11,000 gallons of wine annually, including both varietals (Ravat Blanc, Seyval Blanc) and generics (Barn Red, Waramaug White, and Lakeside White); several are gold-medal winners. The winery also produces 2,000 gallons of cider annually.

Open all year. May-December, daily 11-5; January-May, weekends only, 11-5.

Tours, guided and self-guided; guided by appointment (two weeks notice suggested). Slide presentation. Free tasting. Retail outlet: wine, gifts. Cookbook: Recipe Ideas from Hopkins Vineyard. *Self-guided vineyard walk. Picnic facilities. Access to tasting room and upper level of winery for the handicapped. Directions: From Litchfield, Route 22 west to New Preston. At New Preston, north on 45 past Lake Waramaug to Lake Road. Left on Lake Road to Hopkins Road and winery.*

North Stonington

Crosswoods Vineyards, Inc.
75 Chester Maine Road
North Stonington, Connecticut 06359
Telephone: (203) 535-2205
Owners: Hugh P. and Susan H. Connell

On clear days you can see the Atlantic Ocean and Montauk Point from Crosswoods' hilltop location and surrounding thirty-acre vinifera vineyard. But the distant view is not the only thing worthy of your attention. The Crosswoods winery, a replica of the four-barn 19th-century dairy complex that once stood on the property, is equally deserving. The interior, in dramatic contrast to the old-fashioned facade, has vaulted ceilings and mahogany paneling above its ceramic tile walls and floors. The space is filled with ultra-modern equipment, with an abundance of shiny stainless steel in evidence. Some German ovals for aging wine lend a touch of the traditional.

Tours of this state-of-the-art winery are by appointment only. And while no tastings of its small production (5,000 cases annually) are offered, Crosswoods is worth a stop if you are in the area.

The winery produces mostly estate-bottled vintage-dated varietal wines from grapes grown in its vineyards. These are planted to Chardonnay, Johannisberg Riesling, and Gewürztraminer. To date, the only wine released is Gamay Beaujolais Nouveau; a generic, Scrimshaw White, is promised for the near future.

Open by appointment only. Guided tours by appointment only (24 hours notice), Monday-Friday 9-4; Saturday 9-5. Retail outlet: wine. Access for the handicapped. Directions: Route 195 to Route 2. Route 2 north into North Stonington Village; winery and vineyards one mile north.

Pomfret

Hamlet Hill Vineyards
Route 101
Pomfret, Connecticut 06258
Telephone: (203) 928-5550
Owner: August W. Loos

Hamlet Hill is strategically located near the well-traveled path between two of New England's most popular tourist attractions: Mystic Seaport and Sturbridge Village. The winery is also one of the area's most starkly modern; its domed twelve-sided central building incorporates a combination of ultra-modern insulation techniques in its design. The result is a winery which requires no mechanical heating or cooling devices for the maturation of its wines.

Tours of Hamlet Hill begin with a very complete video presentation which explains every phase of the winemaking process from the harvest and crush to fermentation and aging in oak barrels, to bottling. After this visual explanation, you are free to watch the processes from an interior observation deck; a recorded commentary describes the scene below. Afterwards you can stroll along a walkway through the nearby experimental vineyard, where several different varieties of grapes are being tested.

Beyond the experimental vineyards is Hamlet Hill's producing acreage, planted to European and European-American hybrids including Riesling, Chardonnay, Pinot Noir, Gamay, Gewürztraminer, Seyval Blanc, Chancellor, Cascade, Aurora, and Maréchal Foch. You can sample Hamlet Hill's products at a stand-up tasting bar in the windowed second story of the winery. Hamlet Hill produces the varietal Seyval Blanc, and generics Charter Oak Red, Woodstock White, White Reel, Renaissance Rose, and Pomfret Red. Its Seyval Blanc Extra Dry Reserve is an award winner; Charter Oak Red 1982 and Renaissance Rose are notable, as well. Hamlet Hill is also the exclusive producer of Brunonian Reserve, the private-label wine of Brown University, the owner's alma mater.

Open all year: daily 10-6. Self-guided tours and vineyard walk. Video and audio presentations. Free tasting. Retail outlet: wine and gifts. Picnic facilities in the vineyards. Access for the handicapped. Directions: From New London: 395 north to 169. 169 North to 101. West on 101 to Hamlet Hill (on your left).

MASSACHUSETTS

Bolton

Nashoba Valley Winery
100 Wattaquadoc Hill Road
Bolton, Massachusetts 01740
Telephone: (617) 779-5521
Owner: John F. Partridge, Jr.

A fifty-acre apple orchard, rather than a vineyard, surrounds Nashoba Valley's new winery in Bolton. But an apple crop suits this winery just fine, for Nashoba Valley is the only one in New England to specialize exclusively in the production of fruit wines—premium fruit wines, in fact, that have received many awards in wine competitions.

Nashoba Valley Winery was founded in 1979 by John Partridge in the kitchen and basement of an apartment building where he once lived. A Yale graduate, Partridge became interested in the history of fermented beverages in this country. And while many oenophiles still consider fruit wines the "poor relations" of grape wines, Partridge discovered that the former played a substantial part in early American drinking habits. "In Colonial America nearly every fruit, and occasionally even the vegetables, were made into wine," he says, naming such interesting and even unlikely varieties as black current, gooseberry, ginger, blue fig, turnip, and rhubarb. Thus Nashoba Valley was born—to revive American interest in these delicious but largely forgotten wines.

In five years, the business outgrew Partridge's basement location and a second, larger one, in West Concord. The winery's new home is a custom-built barn designed in a style consonant with the Colonial architecture of New England.

In addition to apples, the surrounding orchard now contains peach, cherry, and plum trees; elderberry, raspberry, and blackberry bushes are planned as well. The remaining fruits used in production are purchased, but are all native to New England; the wild blueberries come from Maine, the cranberries from Cape Cod.

An informative half-hour guided tour of the winery covers the complete winemaking process and reveals some of the special adaptations in equipment and method required to produce wine from fruit other than grapes. These include the equipment used for pitting peaches, chopping fruit, and spinning the flesh in a tank press to extract the juice, as well as adding sugar to complete the fermentation process. You will then see the aging cellar (the drier wines are aged in oak) and the bottling line with its labeling, corking, and sealing processes.

After the guided tour of the working winery, you are free to take a walk through the orchard along a path marked with informative signposts. The acreage is planted with many rare antique apple varieties, including Winter Banana, Westfield Seek-No-Further, and Roxbury Russet. If you visit in the right season (September to November), you might opt to pick your own apples from the less exotic McIntosh, Baldwin, Cortland, and Red Delicious trees and later enjoy a picnic lunch under their boughs at one of the many tables provided for the purpose.

Nashoba Valley produces 24,000 gallons of fruit wine annually. Its product list includes the varietal wine, Winter Banana, and the generics apple, pear, After Dinner peach, a brut style *methode champenoise* sparkling apple wine, cranberry apple, and various berry wines. The winery is particularly noted for its semi-sweet blueberry, its dry blueberry, and its After Dinner peach, all gold-medal winners.

Open all year: daily 11-6. Guided tours of winery: Friday, Saturday, and Sunday, 11-6. Self-guided tour of orchard. Free tasting. Retail outlet: fruit wines, gifts, apples. Pick your own apples: September-November. Picnic facilities throughout the orchard. Access for the handicapped. Directions: From Boston: Interstate 495 to exit 27. West on Route 117 one mile to yellow light (Wattaquadock Hill Road). Left on Wattaquadock, up the hill to winery (¼ mile).

Plymouth

Commonwealth Winery
Lothrop Street
Plymouth, Massachusetts 02360
Telephone: (617) 746-4940
Winemaker and Founder: David Tower

A visit to Plymouth, Massachusetts, to see the remains of the famous landing rock and the living-history *Mayflower II* museum ship is an obligatory pilgrimage for families with young children. But what the adults intent on educating their offspring should also know is that just around the corner from these Water Street historic sites is the Commonwealth Winery.

A tour of the facility and tasting of its product will round out the day's activities.

Commonwealth Winery was founded in 1978 by David Tower, a math teacher turned winemaker, who trained for his current profession by earning a master's degree in oenology at UC Davis and apprenticing in Germany's Rhine Valley for two years. Commonwealth does not own any vineyards; instead, it purchases all of the grapes needed for production from sixteen Massachusetts vineyards, some of which were planted under Mr. Tower's direction.

The thirty-minute tour of the winery begins with a video presentation covering grape growing, harvesting, crushing, pressing, and other elements of wine production. Visitors are then guided through the fermenting area (upstairs) and the barrel room (downstairs.) A detour, across the parking lot to the warehouse, is taken on days when the bottling line, located there, is operating. (There is no set schedule for bottling activities, but this process is frequently operational during summer months.) Sampling of Commonwealth's product follows the tour in a spacious tasting room furnished with two stand-up bars. Antique presses—apple, grape, and basket—as well as an antique stemmer-crusher, add considerable atmosphere to the otherwise matter-of-fact architecture of the vintage-1915 cinderblock warehouse. (Among its other industrial uses before it became a winery, the building served as a bootleg distillery during Prohibition.) Usually four wines are available for tasting, one of which is always the winery's popular and only fruit wine: cranberry apple.

Commonwealth's product line is quite diversified for its 15,000-gallon annual output. It includes the aforementioned cranberry apple wine; five proprietary blends, some of which are named for nearby Plymouth Rock (Plymouth Rock Rosé, Plymouth Rock White, etc.); and varietal wines made from such European varietals as Chardonnay and Riesling, and such French-American hybrids as de Chaunac and Vidal Blanc. Commonwealth is particularly known for its award-winning Quail Hill Chardonnay, de Chaunac Winemaker's Reserve, and Cayuga.

Open all year. April 1-December 31: Monday-Saturday 10-5, Sunday 12-5; January 2-March 31: Monday-Friday 11-4, Saturday 10-5, Sunday 12-5. Guided tours. Free tasting. Retail Outlet: wine and novelty wine items. Access for the handicapped. Directions: Plymouth is an hour's drive south of Boston on Route 3. Winery located on Lothrop Street off Route 3A (Route 3A is called Court Street locally) in downtown area.

West Tisbury

Chicāma Vineyards
Stoney Hill Road
West Tisbury, Massachusetts 02575
Telephone: (617) 693-0309
Owners: Catherine and George Mathiesen

Grapes have always grown wild on the island of Martha's Vineyard, a fashionable summer resort six miles off the coast of Massachusetts. But now the island also has a thriving commercial vineyard to justify its name—Chicāma Vineyards, founded in 1971 by Cathy and George Mathiesen. The moderating climate of the Gulf Stream is favorable to the Mathiesen's enterprise, and visits to the winery are very appealing to the island's many sophisticated vacationers—especially on days when they've had too much sun or the weather isn't cooperating with beach activities.

Chicāma can be found just outside the harbor town of Vineyard Haven on State Road in West Tisbury. Here, about two miles beyond the village limits, a sign directs you down a long, bumpy dirt access road through scrub pines and sandy soil to the vineyard and the winery door. Chicāma is a family-run operation; the Mathiesens have six children who lend their parents a helping hand. Some have apprenticed at other wineries, like daughter Lynn who worked at Domaine Chandon in the Napa Valley; and others, like Timothy, learned the art of viticulture and winemaking from their parents.

Chicāma is a small winery with an annual production of 60,000 bottles, of which eighty-five percent is still table wine and fifteen percent sparkling wine. Thirty-five acres surrounding the attractive, weathered winery structure are planted to fourteen varieties of vinifera grapes. Some of these varieties are still experimental, but others, including Chardonnay, Pinot Noir, Cabernet Sauvignon, and Merlot, are fermented for their current product, supplemented with grapes purchased from as far away as California.

Tours of Chicama are very informative and friendly, probably because they are often guided by an enthusiastic family member. Lasting about thirty minutes, tours cover the winemaking equipment, aging cellar, and bottling line, and end up in the winery's tasting/retail sales area. Here your guide will pour samples of Chicāma's wine and comment on each selection. Afterwards you can purchase bottles of the varietals you particularly enjoyed, all of which are conveniently displayed nearby. Recently, the winery has produced an outstanding Zinfandel (Special Reserve) and Cabernet Sauvignon. Its Chenin Blanc is also very popular, as are the herb and berry vinegars put up in wine bottles by Cathy Mathiesen.

Open daily May 15-October 30: Monday-Saturday, 11-5; Sunday, 1-5. Guided tours. Free tasting. Retail outlet: wine, gifts. Christmas Shop, November 21-December 24: 11-4. Directions: From Vineyard Haven: Southwest two miles on State Road to winery entrance road (on your left).

RHODE ISLAND

Little Compton

Sakonnet Vineyards
West Main Road (Box 572)
Little Compton, Rhode Island 02837
Telephone: (401) 635-4356
Owners: Jim and Lolly Mitchell

Sakonnet Vineyards, Rhode Island's first commercial winery and vineyard since Prohibition, is located in the southeastern corner of the state just outside the lovely seaside town of Little Compton. The winery, which straddles a tract of land lying between the Sakonnet River and the Patchet Reservoir, was founded in 1975 by Jim and Lolly Mitchell. The Mitchells left busy careers in chemical engineering and public relations to start a business that would enable them to spend more time together. Confirmed Easterners, they chose their Rhode Island location after careful considera-tion of both market and climate: Nearby Boston is the third largest wine market in the United States; southeastern New England, recently designated as a viticultural district, enjoys a moderate seaside climate and a growing season similar to that of Burgundy.
    Jim Mitchell's interest in winemaking developed of necessity while he was on assignment in Libya. In order to enjoy his favorite beverage with dinner, Jim had to make it himself, as no spirits are sold in that country. His former hobby provided the inspiration for the new business. Lolly Mitchell, a graduate of Harvard Business School, promotes the results, which are overseen by winemaker Blair Tatman, who graduated from the Culinary Institute of America and trained at UC Davis.
    The winery makes its home in an attractive rambling new building that

contains the working winery, the aging cellar, tasting room, warehouse space, and even living quarters for the Mitchells. Tours of the winery begin with a walk through its surrounding vineyard. Here, forty-five acres are planted to the white grape varietals Riesling, Chardonnay, Pinot Noir, Seyval, Vidal, and Aurora; and to the reds—Foch, Millot, and Chancellor. This harvest is supplemented with purchased grapes to produce Sakonnet's 10,000-case annual output.

After the vineyard walk (which is optional, but encouraged) visitors are guided through the winery. Here are the crush platform, fermentation room, barrel aging cellar, and bottling line. The tour ends with a complimentary tasting.

Sakonnet's product line includes the varietal wines Riesling, Chardonnay, Pinot Noir, and Vidal Blanc, as well as the generics Compass Rose, America's Cup White, Spinnaker White, and Rhode Island Red, appropriately named to reflect their native state. The winery's Pinot Noir (aged in French oak), America's Cup White, and Rhode Island Red are the best known.

The winery has picnic facilities for thirty-six people. It also sponsors numerous festivals and special events throughout the year.

Open all year except major holidays: Tuesday-Saturday, 10-5. Guided tours: mid-May-October, Wednesday and Saturday at 10 and 5. (Winery open off season for tours and tasting by appointment.) Free tasting. Retail outlet: wine. Picnic facilities (food available nearby). Special events: Harvest Vendange (Fall) and Sakonnet Chefs' Series and Cooking Camp. Access for the handicapped. Directions: From Tiverton intersection of Routes 77 and 24, go south on 77 for 8 miles to winery.

Prudence Island

Prudence Island Vineyards, Inc.
Sunset Hill Farm
Prudence Island, Rhode Island 02872
Telephone: (401) 683-2452
Owners: The Bacon Family

It takes a bit of doing to get to isolated Prudence Island Vineyards, located on a small island in Narragansett Bay, and once there, you're at the mercy of the ferry schedule. So plan to make a day of it. Bring your bike, fishing poles, wear a bathing suit under your clothes, and pack a picnic if you wish. For after your visit to the winery and vineyard, you'll find that

Prudence Island is blessed with two state parks, beautiful beaches, a good restaurant (The Prudence Inn, specializing in seafood), a sparse population, and even sparser visitor facilities. (There is no bathhouse, for instance.)

The Prudence Island ferry leaves regularly in summer months from Church Street Wharf in Bristol; the pleasant crossing takes about half an hour. (If you travel by private boat, you can tie up at the island's Homestead Pier.) Most visitors to the winery simply walk up the hill and through the vineyards to the winery's 1783 farmhouse, but taxis are available at the pier for those who prefer to ride.

Prudence Island Vineyards is very much a family-run operation. Bill Bacon, Sr., is the winemaker and his son Nate the vineyard manager; other members of the family take an active part as well. The 400-acre tract, which contains the vineyard, farmhouse, and winery, has been in Natalie Bacon's family for generations. The land was farmed continuously until 1938, when her father turned to other pursuits. But in 1972, Natalie, her husband, Bill, and their two sons, Nate and Bill, Jr., decided to return the acreage to agricultural endeavors once more. They moved from the mainland to the farm, cleared brush and woods from the overgrown land, and planted an all-vinifera vineyard on 16 acres. Since then, they have built a winery in a coverted cow-shed, and even blasted and poured, with little outside help, a 14,440-cubic-foot underground wine cellar.

Tours of the winery are quite relaxed and are tailored to the interests of the participants. They take about thirty minutes and cover the winery and vineyard, ending with a tasting of the Prudence Island product. All of the grapes for Prudence Island wines are grown in its beautiful vineyard, which is planted to the grape varietals Chardonnay, Gamay Beaujolais, Riesling, Gewürztraminer, Cabernet Sauvignon, and Merlot. These varieties produce 2,500 gallons of table wine annually, most of it Chardonnay and Gamay Beaujolais. The Chardonnay has been awarded both gold and silver medals.

Open daily: Memorial Day-Labor Day, 10-12 and 2-5; otherwise by appointment only (weekends best from 10:30-3:30). Guided tours daily, Memorial Day-Labor Day, or by appointment. Free tasting. Retail outlet: wine, winemaking supplies, gifts. Picnicking permitted on farm and its private beaches; state parks nearby. Access for the handicapped. Directions: From Bristol, take ferry departing from Church Street Wharf (off Thames Street) to Prudence Island's Homestead Pier (½ hour trip). From Homestead Pier, either take a taxi, or walk up the hill to the end of Pier Road. Turn left on gravel road and follow it through the vineyards to the Bacons' farmhouse. (The walk from the ferry to the winery takes about 10-15 minutes.) Call winery for ferry schedule. In summer months it generally leaves Bristol at 8, 10, 3:30, and 6.

 New York State

Since the repeal of Prohibition, New York has led the states east of the Rockies in vinous endeavors. In fact, with 76 bonded wineries, 42,000 acres of vineyards, and 32 million gallons of wine produced annually, it is second in the nation only to California. Wineries exist in most parts of the state—from Long Island to the western Chatauqua-Erie belt along Lake Erie. For the traveler planning a tour, these have been divided for convenience into three large regions—the Finger Lakes and Western New York, the Hudson River Valley, and Long Island.

The greatest concentration of wineries and New York's largest facilities are found in the Finger Lakes region. Here one finds such large firms as Taylor/Gold Seal/Great Western, Canandaigua, and Widmer. Clustered around the shores of Seneca, Keuka, and Cayuga Lakes, the wineries of the area are most hospitable, vying with one another for their share of the tourist traffic with video presentations on winemaking, hayrides through the vineyards, open-air restaurants, music festivals, marathon and boat races, museums of wine antiques and history, and handsome landmark buildings turned into hospitality centers. In addition to its wineries, the area is also rich in scenic beauty and other tourist attractions such as the Corning Glass Center, Watkins Glen, and numerous state parks that provide facilities for fishing, camping, and boating.

The Hudson River Valley wine region, within easy reach of the New York metropolitan area, lays claim to being the oldest wine district in the state. Its Brotherhood Winery, founded in 1839, was the second winery to be bonded in the United States. And Benmarl, the state's first farm winery, located a little further up the Hudson, can trace back plantings on its estate to the early 1800s. About half of the Hudson Valley's vineyards are planted to American varieties while the other half is roughly divided between French-American hybrids and vinifera.

New York's newest and smallest winegrowing district—many feel it is the state's most promising—is on the eastern end of Long Island. Farms used for generations for growing potatoes are being bought up and the humble barns and cellars converted to showplace wineries. Prize-winning wines made largely from vinifera grapes are being produced. Long Island's climate, moderated by the surrounding ocean and a 210-day growing season (45 days longer than that of the Finger Lakes) has always seemed ideal for growing viniferas, but, until the development of pesticides to combat the insects that thrive in this humid region, the planting of commercial vineyards was not feasible. Three wineries have been bonded on Long Island's north and south forks since the mid-1970s. Licenses for two more are pending.

Two events have indelibly changed the course of New York State winemaking in recent times and will shape its future. One was the 1976 passage by the state assembly of legislation favorable to the establishment of farm wineries; the second was the arrival of Russian-born Konstantin Frank, a private winegrower who demonstrated to his colleagues that viniferas can thrive in the state. As a result of these events and the growing demand for more sophisticated wines by the public, the state's dominant Concord plantings have been slowly giving way to the less harsh Labrusca varieties such as Dutchess, Delaware, and Catawba, as well as French-American hybrids and viniferas. Winegrowing, rather than grapegrowing, is gradually becoming the focus of the industry.

THE HUDSON VALLEY

Amenia

Cascade Mountain Vineyards
Flint Hill Road
Amenia, New York 12501
Telephone: (914) 373-9021
Owners: Bill and Margaret Wetmore

Bill Wetmore can trace his family's grape-growing and winemaking roots back to the late 19th century. In 1882, a distant relative, Charles Wetmore, planted the famous Cresta Blanca Vineyards in California. His first vintages, produced seven years later, won him the coveted Grand Prix at the Paris Exposition.

Ninety years and three thousand miles later, novelist Bill Wetmore and his wife, Margaret, planted Cascade Mountain's fifteen-acre vineyard in New York's Dutchess County. Six years later, they took their first vintages to "Wineries Unlimited," the largest international wine competition in the East, and won two firsts.

Cascade Mountain is a family-owned-and-operated winery with the usual amenities for visitors: guided tours, free tastings, and seasonal festivals. In addition, however, there are some special extra touches, the most interesting being an unusual restaurant/barbecue facility. The Wetmores will provide a lavish repast (for $5 per person), which can be enjoyed either indoors or out. The meal includes a choice of two patés, two soups (usually gazpacho and vichyssoise), smoked salmon or trout, salad, fresh fruit, and pastries. If you prefer a hot entrée, the Wetmores will substitute veal, local beef, or lamb, which you can grill to your taste over a charcoal fire stoked with dried vine cuttings.

Another special touch is the rural splendor of the winery's setting: it is surrounded by seventy acres of vineyard, pasture, woods, and ponds which guests are free to explore and enjoy.

Cascade Mountain produces 5,000 cases of wine annually. The grape varieties Aurora, Leon Millot, Maréchal Foch, Chancellor, and Baco Noir are planted on the southern slope of the mountain which gives the vineyard its name. From these grapes (and additional ones purchased) the winery produces the varietal table wines Chancellor and Aurora, a number of proprietary wines with such fanciful names as Le Hamburger Red and Spring Fever, and a small quantity of fruit wine. Le Hamburger Red and Little White Wine, both dry table wines, are popular as well as inexpensive.

Open all year: daily 10-6. Tours: guided for groups, self-guided for individuals. Free tasting. Retail outlet: wine. Restaurant. Picnic facilities and barbecue. Festivals: spring and fall. Directions: From light in Amenia go north on Route 22 3 miles. Follow signs to winery.

Highland

Hudson Valley Wine Co., Inc.
Blue Point Road
Highland, New York 12528
Telephone: (914) 691-7296
Owner: Herbert Feinberg

The appropriately named Hudson Valley Wine Company is located in the heart of New York's scenic Hudson River valley on a 310-acre estate with nearly a mile of river frontage, three ponds, hiking trails, an apple orchard, and a 100-acre vineyard. The winery complex, located on a bluff overlooking the Mid-Hudson Bridge and the village of Poughkeepsie across the river, is equal to its handsome setting. Established in 1901 by Adolpho Bolognesi, a prominent New York City investment broker, the complex is made up of a cluster of picturesque stone buildings with decorative arches and other masonry flourishes created by the Italian stonemasons Bolognesi employed.

Hudson Valley Wine's buildings include the press room, with its crusher/destemmer, press, and various types of cooperage; the combined operating facility and wine cellars, enhanced by an old clock tower; the Champagne Room, housing the high-pressure champagne tanks, the winery's laboratory, and a museum; and the manor house. The last, the former summer home of the Bolognesis, is a fourteen-room mansion with rich woodwork, stained-glass windows, and a rear veranda offering a spectacular view of the river and the valley. The manor house now serves as the tour reception center; its dining room is used for catered buffets.

The winery's vineyards, some tended by descendants of the original caretakers of the property, are planted primarily to Catawba; other varieties include Delaware, Iona, Concord, Chelois, and Baco Noir. From these grapes are produced 50,000 cases of wine annually—sixty percent table wine, thirty percent sparkling wine, and ten percent fruit wine. The product line includes more than a dozen different wines: native Labruscas, French hybrids, champagnes, and fruit wines. The winery is particularly known for its Hudson Valley White Burgundy, Pink Catawba, and Hot Rumour (a mulled spiced wine).

The old Bolognesi estate, once fenced and guarded to assure its owners' privacy, now attracts thousands of visitors each year. Comprehensive one-hour tours of this winemaking "village" encompass the working winery buildings; tours end in the wine shop and tasting room, where you are presented with a souvenir wineglass (on weekends only), are encouraged to snack on cheese and bread as you sample the Hudson Valley product, and can purchase wine to take home.

Hudson Valley Winery hosts a special activity, festival, or event almost every weekend from mid-March to the end of December. Among the festivities are an annual Easter egg and champagne hunt, a strawberry festival, a fishing contest, watermelon festival, North American Grape Stomping Championship, Pick-Your-Own-Grapes Weekend, and much more, including hayrides for the children. (Write for a current schedule.)

Open all year, 9-5: weekends, December-March; daily, April 2-November 2. Weekend admission: $5 adults, $1 children; weekdays: $3 adults, children free. For special events add $1 per person. Tasting, cheese, bread sticks, and souvenir wineglass (weekends only) included in entry fee. Also included: guided tours, regularly scheduled throughout the day. Museum. Retail outlet: wine. Snack bar. Picnic facilities. Festivals and special events every weekend from March-December. Partial access for the handicapped. Directions: New York State Thruway to exit 18 (New Paltz). Route 299 east to 9W. Take 9W south for 3½ miles to winery.

Marlboro

Benmarl Vineyards
Highland Avenue
Marlboro, New York 12542
Telephone: (914) 236-7271
Owners: Mark and Dene Miller

At Benmarl, tours and tastings are by appointment only and there is also a charge for the latter. But if you go to the trouble of planning ahead, you won't be disappointed by your visit here. For Benmarl is one of the East's most attractive wine estates and produces some of its finest wines.

Mark and Dene Miller established their winery on a slate ridge (the Gaelic word *Benmarl* means "slate hill") outside the village of Marlboro on land first planted to vineyards in the early 1800s. The Millers replanted the acreage to vineferas and French-American hybrids in 1957; their 15,000-case winery was bonded fourteen years later.

Benmarl's wood and brick buildings were designed by Dene Miller, an architectural draftswoman. Her husband, Mark, an artist by profession, tends the vines and makes the wine. (His artistic talent is reflected in the winery's special museum of wine-related sculpture and drawings.)

Many of the seventy-five acres planted to vineyard are owned by the Société des Vignerons, a cooperative association the Millers formed to sell vine rights to individuals (the initial fee and annual dues entitle members to a quantity of wine and to other privileges each year. The Société has helped to finance the winery's expansion while giving serious oenophiles a share in a working vineyard and winery. (The idea of this cooperative venture, imported from Burgundy, has since been repeated elsewhere in this country.)

Guided tours of Benmarl focus on the seasonal activities of both vineyard and winery and cover the entire operating facility, as well as the museum. There is a fine picnic site on the property and, from the uppermost ridge of the vineyard, a spectacular view of the Hudson River and valley below.

Benmarl produces only premium table wines, which in most cases are blends of different varietals. It is known for its long-lived dry reds and subtle dry whites. Until recently, its prize-winning wines were completely sold out to Société members each year; increased production now permits wider distribution.

Open all year: by appointment only. Guided tours. Tasting (nominal charge). Retail outlet: wine, gifts, books, artwork. Art Museum. Picnic facilities. Société des Vignerons: membership $500, annual fee $75. Directions: winery will supply when appointment is made.

Milton

Royal Kedem Winery
Route 9W (P.O. Box 811)
Milton, New York 12547
Telephone: (914) 795-2240
Owners: The Herzog Family

A popular wine touring spot, Royal Kedem Winery attracts thousands of visitors each season. A long-established Eastern European winery, Royal Kedem was re-founded in Brooklyn, New York, in 1948, and has been run by the same family for seven generations (the Herzogs emigrated to the United States in that year from Czechoslovakia). Although expanded tremendously since the early days, and now located in the Hudson River town of Milton, Royal Kedem is still family-owned and operated, and still makes its strictly Kosher wines in accordance with old recipes and traditional techniques handed down from one generation to the next.

The hour-long guided tour of the winery includes the working facility, where winemaking and cooking with wine are explained; a film covers the seasonal harvest, the crush, and other steps in processing. At the tour's conclusion, you may sample as many as twelve Kedem wines at a horseshoe-shaped tasting bar in what was once Milton's train station.

The winery purchases 4,000 tons of grapes annually, both from local growers and from the Finger Lakes region of the state. Royal Kedem's product line is one of the most extensive of the Hudson River wineries. Its staggering one-million-gallon annual production includes a broad range of wines, from a dry Seyval Blanc and Chablis to sweet Honey and Plum. It also produces a *methode champenoise* sparkling wine. Particularly noteworthy are its Chablis, Seyval Blanc, and White Champagne.

Open all year, except Saturdays and Jewish holidays: May 1-December 31, Sunday-Friday, 10-5; January 1-May 1, Sundays only, 10-5. Guided tours. Film. Free tasting. Retail outlet: wine, gifts. Festivals: May and November. Picnic tables overlooking the Hudson. Access for the handicapped. ($1 parking fee, Sundays only.) Directions: New York State Thruway north to exit 17 (Newburgh). From Newburgh follow 9W north to Milton center (9 miles). At blinking light in town center turn east and follow signs to winery.

Washingtonville

Brotherhood Winery
35 North Street
Washingtonville, New York 10992
Telephone: (914) 496-3661
Owner: E. D. Farrell

Brotherhood Winery, with its beautiful fieldstone buildings and historic underground cellars set in the rolling foothills of the Catskills, makes a popular day's outing from New York City and environs. One of the oldest bonded wineries in the United States, Brotherhood no longer grows its own grapes; its once expansive vineyard has given way to visitor parking. But the winery's informative guided tour and structured tasting, lasting nearly ninety minutes, help to compensate for the lack of an estate vineyard, as do the manicured lawns, old gardens, and secluded picnic sites set amid groves of trees. The place is even charming in winter: December is an especially festive month to visit; the cellars are decorated and lit for the holidays, and hot mulled wine, eggnog, and cheese and crackers are served.

Brotherhood Winery was established in 1810, when Jean Jacques, a French emigré, planted a vineyard and built a small winery in Washingtonville. An elder in the Presbyterian Church, Jean preferred to sell his wine for sacramental purposes. But word of his product spread, and before long he was doing a lively business with the local gentry, who bought the wine under the pretext of its medicinal value. As the vineyard prospered, Jean's sons became convinced that it would be wise to share the bounty with the public at large and began selling their wine to Jesse M. Emerson, a New York City wine merchant. Emerson also bought wine from a religious community, called Brotherhood, across the Hudson. And before the end of the century, he bought the Jacqueses' family winery, adopting the Brotherhood name after the religious community moved westward. He constructed the handsome fieldstone buildings which still exist, and dug the now-famous underground cellars, while developing a thriving market for the wine throughout the United States, Europe, and Africa.

Brotherhood survived Prohibition by reverting to its first business—making wine for churches. Shortly before Repeal, controlling interest in the business was acquired by Mr. L. L. Farrell; his wife and daughter own and operate the winery today.

Tours of Brotherhood begin with a glass of wine; sip while your guide gives a short talk on the history of winemaking and the enjoyment and

appreciation of wine. Next is a short slide presentation, followed by a visit to the fermentation vault, with its large oak casks, and the underground cellars. The aging cellars, purported to be the largest and oldest in the country, are lined with handmade casks in which wine has been aged for generations. Several different cellars are included on the tour so that your guide can explain the aging process and how it varies from one wine to another. At the conclusion of the tour, you can taste a variety of Brotherhood's twenty wines while your guide comments on how each should be served and stored, and with which foods it is best enjoyed.

Brotherhood makes its wine from grapes purchased from many of the grape-growing regions of New York State. Its product line, the recipient of many awards, includes table, fortified, and aromatic wines. Brotherhood is particularly known for its May Wine, Rosario, sherries, and ports.

Open February-December, except Good Friday and Easter: Call or write for varying times and days open. Guided tours (regularly scheduled), adults $2. Retail outlet: wine, gifts. Picnic facilities. Festivals. Wine tasting workshops. Directions: New York State Thruway to exit 16 (Harriman). Take 17 west to exit 130. Follow Route 208 north to Washington and the winery.

THE FINGER LAKES AND WESTERN NEW YORK

Bluff Point

Chateau Esperanza Winery
Route 57A (P.O. Box 76)
Bluff Point, New York 14478
Telephone: (315) 536-7481
Owner: Angela Lombardi

Chateau Esperanza, located on a high bluff overlooking Keuka Lake, is prominent even from a distance. The winery's home is a twenty-room Greek Revival mansion, built in 1823, which originally served as the country residence of a wealthy family. In 1979 the building began its new life as a winery and is currently being restored to its once handsome state. When completed, it will provide a suitable setting for Chateau Esperanza's wines, which though the product of a young winery, have already garnered over thirty medals, several of which were best-of-competition awards.

Chateau Esperanza is committed to the production of premium wines made from vinifera and French-American hybrids. Currently these wines are made from purchased grapes, as the winery's young seven-acre vineyard is not yet producing. Wines offered range from dry to dessert; all are vineyard designated. Included in the product line are a Late Harvest Ravat, Late Harvest Johannisberg Riesling, and Chancellor Noir as well as the generics Chateau White and Chateau Red. A *methode champenoise* sparkling wine is planned for the near future. Particularly notable among the prize-winning wines is the 1982 Ravat, Late Harvest, which was awarded both Gold Medal and Best of Class in the 1983 Eastern Wine Competition. The 1984 Chancellor Noir is also recommended.

Weekday tours of the winery are self-guided and consist of the restored main-floor rooms, which house the tasting area, retail outlet, and a museum displaying both old and new winemaking equipment and posters outlining the history of the house and its occupants. On weekends, guides are on hand to explain the working winery, which is located in the cellars down a flight of steep, narrow stairs. In addition to reviewing the technical aspects of winemaking, you'll learn about the cellar's past, including its use as a way station on the underground railroad during the Civil War.

Picnicking facilities are provided on the expansive front lawn, with an unequalled vista of the lake and hillsides beyond. Directly opposite the winery, on the east side of Route 54A, is the entrance to Keuka Lake State Park, a major attraction for campers and boaters.

Open daily, May 1-October 31: Monday-Saturday, 10-5; Sunday, 12-5. Guided tours of winery, weekends only. Self-guided tours of house at all times. Museum. Free tasting. Retail outlet: wine, gifts. Picnic facilities. Access for the handicapped limited to first floor of mansion. Directions: From Interstate 90 take exit 42. Follow Route 14A through Geneva to 54A, then west on 54A to winery.

Branchport

Finger Lakes Wine Cellars
Italy Hill Road (R.D. 1, Box 13)
Branchport, New York 14418
Telephone: (315) 595-2812
Owners: Arthur and Joyce Hunt

On the northwestern tip of Keuka Lake, west of Chateau Esperanza on Route 54A, you will come to Finger Lakes Wine Cellars, the rustic farm winery of Arthur and Joyce Hunt. Founded in 1973 on land owned and farmed by the Hunt family for six generations, the winery is housed in a converted 1820 barn. Beyond is a seventy-acre vineyard planted to native American grapes, French-American hybrids, and viniferas: Riesling, Chardonnay, Ravat, Seyval Blanc, Cayuga, Chancellor, Aurora, Delaware, de Chaunac, Niagara, and Concord.

From this crop and additional purchased grapes (mainly viniferas), the Hunts produce 10,000 gallons of wine annually. Their product line—made up solely of still table wines—includes varietals of the aforementioned grapes and the generics Classic White and Classic Red. To date these wines have only been available at the winery or at select local restaurants and outlets. Award winners include Riesling, Chardonnay, Late Harvest Ravat, Cayuga, and Delaware.

Finger Lakes Wine Cellars is a particularly hospitable place to visit. There are hayrides through the vineyards to observe their meticulous cultivation at first hand; this trip is supplemented by a slide presentation that covers the development of grape vines and the maturation of the grapes, the history of the winery, and winemaking. Tastings conclude your visit. In nice weather, the wines are sampled on the deck off the tasting/retail room in view of the vineyards. And at harvest time, fresh grape juice and grapes are available.

Open May 1-October 31: Tuesday-Saturday, 10-5; Sunday, 1-4. Guided tours. Hayrides through the vineyards (adults $1.50, children under 10 free). Slide presentation. Tasting ($1 charge refunded with purchase). Retail outlet: wine, gifts. Picnic facilities. Limited access for the handicapped. Directions: New York Thruway to exit 42. South on Route 14 to Dresden, west on Route 54 to Branchport. Winery one mile up the hill on Italy Hill Road. Finger Lakes Wine Cellars Visitor Center and tasting room, Route 54, 2 miles south of Hammondsport, open all year, Monday-Saturday, 10-5. (Telephone: [607] 569-2380).

Canandaigua

Canandaigua Wine Company
116 Buffalo Street
Canandaigua, New York 14424
Telephone: (716) 394-3630
Owner: Marvin Sands

Second largest Finger Lakes winery. Twelve-million-gallon plant offers no tours, but operates a tasting room on Route 20 in remodeled century-old building adjoining historic fifty-acre Sonnenberg Gardens. (The combination is the city's chief tourist attraction.) In addition to free samples of the winery product, the firm's handsome tasting room, ornamented with an elaborate stained-glass triptych, includes a display of winery cooperage and artifacts. Open daily, May-October: 12-6. Free tasting. Retail outlet: wine and wine-related merchandise. Picnic facilities. Directions: Located on Route 20 in Canandaigua.

Dundee

Glenora Wine Cellars
Glenora-on-Seneca (R.D. 4, Box 58)
Dundee, New York 14837
Telephone: (607) 243-5512
Owners: Gene Pierce, Eastman Beers, Howard Kimball, Ed Dalrymple, and Bill Thornton.

Glenora Wine Cellars, overlooking the shores of Lake Seneca (the deepest of the Finger Lakes), was founded in 1976 by a group of independent grape growers. The winery building, completed in 1977, is a shingle-sided one-story structure reminiscent of local barn architecture. Its large windows, a concession to modern design, overlook the lake, as does the long deck which extends beyond and also affords a view of the crush pad and pressing facility. In summer months, the expansive winery lawn is dotted with barrel planters filled with flowers; near the vineyards there is a covered picnic pavilion.

Tours of Glenora are limited to a fifteen-minute slide presentation which runs continuously throughout the day, covering the vineyard and winery from harvest to final product. Afterwards, five Glenora wines are available for tasting in a room adjoining the theater.

Glenora produces 35,000 gallons of wine annually. Over 400 acres of estate-owned vineyards are planted to Chardonnay, Seyval, Riesling, Ravat, and Cayuga. From these and additional purchased grapes the winery produces varietal table wines and a small quantity of *methode champenoise* sparkling wine. Glenora's wines, among them its Johannisberg Riesling, Seyval Blanc, and Chardonnay, are consistent prize winners.

The winery owners support local cultural endeavors and host many special events throughout the year, including jazz concerts on Sunday afternoons in the summer months and the Glenora Cup Championship Hobie Cat Races on nearby Lake Seneca in July.

Open all year. May 1-October 31: Monday-Saturday 10-5, Sunday 12-5. November 1-April 30: Monday-Saturday 10-4. Audio-visual presentation; free tasting. Retail outlet: wine, gifts. Picnic facilities. Special events. Access for the handicapped. Directions: New York State Thruway to exit 42. Follow Route 14 south approximately 24 miles to winery (on the west side of the lake).

Dunkirk

Woodbury Vineyards
South Roberts Road
Dunkirk, New York 14048
Telephone: (716) 679-1708
Owners: The Woodbury Family and partners

Woodbury Vineyards, a small (25,000 gallon) premium winery, is located in the heart of the Chautauqua-Erie grape belt, a sixty-mile stretch of land along the shores of Lake Erie. In an area historically planted to Concords, Woodbury is one of a new wave of wineries, founded after the passage

of New York's farm winery legislation in 1976, that are growing vinifera and French hybrids as well. The Woodbury family has owned its farmland, originally planted to orchards, since 1908. (An apple orchard near the winery building, with picnic tables set under its trees, is both a pleasant place to relax and a reminder of that earlier history.)

The working winery is a simple, functional block building with a large, airy cedar-sided addition housing the visitor facilities. Guided tours, regularly scheduled, are available every day. They take about thirty minutes and cover the outdoor press pad, fermentation tanks, wine cellar, bottling line, and champagne cellar, concluding with a tasting of five or six Woodbury wines.

Woodbury Vineyards' product line of award-winning wines, dominated by varietals made from both viniferas and French-American hybrids, is all estate bottled and vintage dated. The list also includes smaller quantities of wine made from native American grapes (Niagara and Dutchess), three proprietary blends (red, white, and rosé), a few fruit wines, and four *methode champenoise* sparkling wines. The winery is particularly noted for its Chardonnay, Johannisberg Riesling, Seyval Blanc, and sparkling wines. Its neighboring vineyards cover forty-five acres at present and are soon to be doubled in size.

Four special events are held during the year, including a May apple blossom festival; a July Fourth celebration; a November harvest festival;

and an unusual winter weekend, called Glacier Ridge, held in February, which includes cross-country skiing, a snow sculpture contest, and horse-drawn sleigh rides.

Open daily all year: Monday-Saturday 10-5, Sunday 12-5. Guided tours. Free tasting. Retail outlet: wine, gifts. Festivals: February, May, July, and November. Picnic facilities. Access for the handicapped. Directions: New York Thruway (Interstate 90) to exit 59. South on Route 60 to Route 20. East on Route 20 to South Roberts Road, right on South Roberts to winery (on your right).

Fairport

Casa Larga Vineyards
2287 Turk Hill Road
Fairport, New York 14450
Telephone: (716) 223-4210
Owners: Andrew and Anne Colaruotolo

Casa Larga Vineyards, only a fifteen-minute drive from downtown Rochester, is a small family-owned operation. Its stuccoed winery's estate vineyard is a green oasis in the suburban landscape of tract homes and commercial real estate which surrounds it. But the location is a hospitable one for grape growing: the soil is stony and well drained, and a constant breeze keeps insects to a minimum.

Casa Larga was founded by Italian-born Andrew Colaruotolo in 1974. A local builder, he first planned to develop the land, but decided instead to plant a vineyard and start a winery, which he named after his family's vineyard in Italy. Casa Larga's seventeen-acre estate vineyard is planted to viniferas (Cabernet Sauvignon, Johannisberg Riesling, Chardonnay, Pinot Noir, and Gewürztraminer) and French-American hybrids. From the former, Casa Larga makes vintage-dated varietals; from the latter, vintage-dated blends such as Casa Larga Red, Casa Larga White, and Rosé.

Tours of the winery begin in the vineyard with a discussion of viticultural practices; include the working winery with its crushing facility, aging cellars, and bottling line: and conclude with a sampling of Casa Larga wines in the tiled tasting room.

Since its first yield in 1978, the winery has collected nearly three dozen awards for its vintages, particularly its Johannisberg Riesling. There are picnic facilities on the grounds, which afford a clear view of the Rochester skyline.

Open all year: Tuesday-Friday 10-5, Saturday 12-5, Sunday 1-5. Guided tours May 1-October 31 at 1, 2, 3, and 4. Free tasting. Retail outlet: wine, winemaking kits, gifts. Picnic facilities. Access for the handicapped. Directions: New York State Thruway to exit 45, then right on Route 96 (towards Victor) to second traffic light. Turn right at light to winding Route 44; follow 44 to the top of the hill. Winery will be on your left. Mailing address: 27 Emerald Hill Circle, Fairport, New York 14450.

Hammondsport

Bully Hill Vineyards
Greyton H. Taylor Memorial Drive
Hammondsport, New York 14840
Telephone: (607) 868-3561
Owner: Walter S. Taylor

Bully Hill Vineyards, set high above Keuka Lake, is located on the original Taylor Wine Company estate. Walter Taylor, grandson of that company's founder, and his father, Greyton, bought back the property from subsequent owners in the 1950s, some thirty years after the Taylor Wine Company had moved its operation to a new location several miles away.

The owner of Bully Hill, Walter Taylor, is an outspoken and colorful figure, known for his championing of French-American hybrids in New York State. (Taylor has no connection with his family's original winery, which is now owned by Seagram, and is described in the following listing.)

Bully Hill is well equipped for visitors. The large estate includes, in addition to a 45,000-case winery and 150 acres planted to French-American

hybrids, a wine museum, winemaker's shop, restaurant, and lodging facilities. Tours of the winery's handsome wood buildings take in the major aspects of the working winery, such as its vineyard, crush pad, fermentation tanks, bottling line, and aging cellars. The museum contains a variety of old wine artifacts collected from other wineries in the Finger Lakes region; the adjacent wine shop sells winemaking supplies and equipment, as well as grape juice. Bully Hill's restaurant, also part of the complex, offers lunch from May through October.

The Bully Hill Vineyards product list includes varietals such as Cayuga, Ravat, Vidal, Seyval Blanc, and Baco Noir, as well as jug wines with such amusing names as Walter S. Who?, Big Bird, and Happy Hen White. It also produces *methode champenoise* sparkling wine, and is particularly known for the latter, as well as for its estate-bottled wines.

Open April 1-December 1: Monday-Saturday, 9:30-4:30; Sunday, 12-4:30. Guided tours daily (weather permitting), every hour or half hour depending on demand. Free tasting. Retail outlet: wine, winemaking supplies. Restaurant: open May-October. Picnic facilities. Greyton H. Taylor Wine Museum ($1 donation suggested). Bully Hill Bed and Breakfast. Special events: ox roasts in July and September feature fireworks, music, and more. Access for the handicapped. Directions: Interstate 390 to Bath exit. North on Route 54 to 54A; left on 54A to winery.

Taylor/Great Western/Gold Seal Winery
Route 88
Hammondsport, New York 14840
Telephone: (607) 569-2111
Owner: Joseph E. Seagram & Sons, Inc.

The most popular stop on any wine lover's itinerary of the Finger Lakes region is the combined Taylor/Great Western/Gold Seal Winery southwest of Hammondsport. Located in the former bottling facility of the Great Western Winery, the hospitality center for these three historic New York wineries clustered near the south end of Keuka Lake has recently been renovated by their parent company, Joseph E. Seagram & Sons.

All three wineries are among the earliest and most important in the area. Great Western, the oldest, was founded in 1860 by twelve local businessmen headed by Charles Davenport Champlin. They built their winery on the slope overlooking Pleasant Valley, calling it the Hammondsport and Pleasant Valley Wine Company. The first bonded winery in New York State, it was renamed Great Western when purchased ninety-five years later by a New Jersey investor. He subsequently sold the company to its neighbor, the Taylor Wine Company. Great Western's wood and stone buildings and cellars, listed on the National Register of Historic Places, are still in use today. The winery is noted both for its champagnes and for premium New York State wines, including vintage varietals, generic table wines, sherries, and ports.

Gold Seal, founded in 1865 as the Urbana Wine Company by a group of local investors, was the second bonded winery in the state. In 1887, Urbana introduced the Gold Seal trademark under the tenure of Jules Crance, a French winemaker who used it to commemorate the company's award-winning wines. The name was not adopted permanently until 1957, however, in recognition of the harvest and production of a bumper crop of experimental winter-hardy viniferas. Gold Seal's product line includes Charles Fournier Special Selection Champagnes, estate-bottled vintage varietals, New York champagnes, generic table wines, and dessert wines.

Established in 1880 by Walter Taylor, a master cooper turned winemaker, the Taylor Wine Company's first home was a seventy-acre farm not far from its present location, the old stone winery buildings of the former Columbia Wine Company. The firm was owned by the Taylor family until the death of the last of Walter's sons in the late 1970s; in 1983, it was acquired by Seagram, which had purchased neighboring Gold Seal five years earlier. Today Taylor produces New York State champagnes, Lake Country and Taylor New York State premium table wines, Lake Country Soft wines, and a popular line of dessert wines.

Visitors to the combined hospitality center for these three wineries are received in the spacious reception building. The center's unique theater, a mammoth 35,000-gallon wine tank now fitted with seats and screen, is located here, as is its wine shop with tasting bar. Either before or after an extensive tour of the winery facilities, you can view a twenty-minute film on the development of the grape industry in Hammondsport. The tour is by bus along a route which leads past the press building to the old stone winery structures further along the road. A guide is on hand to accompany you into specific buildings, including the vineyard room, research lab, and processing building. The tour ends in a mid-19th-century stone fermentation vault which has been converted to a hospitality room; here you can sample six different wines, usually including champagne, a dry white, a dry red, a rosé, a Taylor Soft, and a sherry.

At the close of the tour, you will have an opportunity to taste other of the companies' wines back at the main reception building. You can catch the film if you missed it earlier and examine exhibits of wine memorabilia displayed here, as well. A short walk from the main building is a park, where picnic tables, chairs, and barbecues are set out near a small stream. (On Friday nights in July and August, the park is the setting for a series of concerts.) You'll want to allow at least two hours for your visit here, as there is a great deal to see.

Open all year, except major holidays. January-April: Monday-Saturday, 11-3; May-October: Monday-Sunday, 10-4; November-December: Monday-Saturday 11-3. Guided tours. Film. Free tasting. Retail outlet: wine. Picnic and barbeque facilities. Concerts July and August. Access for the handicapped limited to the reception building. Directions: Route 17 to exit 38 (Bath). From Bath, take Route 54 north for approximately 5 miles before turning left at Pleasant Valley. You'll see signs to the visitor center, located about half a mile beyond.

Hector

Wickham Vineyards
1 Wine Place (P.O. Box 62)
Hector, New York 14841
Telephone: (607) 546-8415
Owners: The Wickham Family

On the east shore of Seneca Lake, on a hill overlooking the village of Hector, is the striking winery of Wickham Vineyards, a modern two-story wood and glass building consisting of two connecting wings. The larger, bordered on three sides by an observation deck, contains a hospitality room for visitors. Tours of the winery begin here with a brief discussion of Wickham's history.

For nearly eighty years, and five generations, the Wickhams have been growing native American grapes in their 165 acres of vineyards for other wineries, but in recent years they have planted an additional ten acres to French-American hybrids, responding to changing trends in the region. In 1981, under the direction of William Wickham V, the family built a winery in the midst of one of their oldest and most fertile vineyards and hired a California-trained winemaker to oversee production.

The tour includes both the outdoor press pad and the indoor production facility, with its stainless-steel fermentation tanks, and aging cellars filled with American oak barrels. It takes about forty-five minutes in all and concludes in the hospitality room for a sampling of five or six Wickham wines.

Wickham currently produces 25,000 gallons of wine annually. All of this is still table wine made primarily from grapes grown in its own vineyards. These are planted to the grape varieties Aurora, Delaware, Seyval,

Ravat, Cayuga White, Baco Noir, de Chaunac, Catawba, and Concord. The resulting product list includes varietals of these, in addition to Chardonnay and Riesling, and two generics: Rose and Mellow Red Table Wine. Wickham is particularly known for its Chardonnay, Baco Noir, Johannisberg Riesling, and Ravat Vignoles.

Picnicking is permitted on the small lawn in front of the winery. Tables and chairs are provided, or you can spread a blanket. There is a good view of one of the family-owned vineyards a few yards away, the village of Hector below, and a glimpse of Seneca Lake in the distance.

Open all year. July 1-October 31: Monday-Saturday, 10-6; Sunday, 1-6; November-June 30: Monday-Saturday, 10-4:30; Sunday, 1-4:30. Guided tours. Free tasting. Retail outlet: wine, gifts. Picnic facilities. Limited access for the handicapped. Directions: New York State Thruway to exit 41 (Route 414). Follow Route 414 south to winery in village of Hector.

Lodi

Wagner Vineyards
Route 414
Lodi, New York 14860
Telephone: (607) 582-6450
Owner: Stanley A. Wagner

Wagner Vineyards, across the lake from Glenora Wine Cellars in Dundee, has all the elements of the perfect wine touring experience: scenic setting, estate vineyards, premium wines, dramatic architecture, full tour, generous tasting policy, and *al fresco* restaurant.

Stanley Wagner, its owner, is one of the vintners who established wineries in the Finger Lakes region after passage of the state's farm winery bill in 1976. A grower for thirty years, Wagner was well equipped to do this. All his life he had been a farmer, and as early as the mid-1950s he began converting his dairy operation to viticulture. By 1965, grapes accounted for almost seventy-five percent of his agricultural efforts; he sold Concords to the Taylor Wine Company in Hammondsport and also began to plant viniferas and French-American hybrids. In 1976, Wagner designed and built his unusual octagonal winery, hired a UC Davis-trained winemaker, and, with the 1978 harvest, completed his first crush.

Since then, Wagner has continued to sell grapes to other wineries, but reserves enough of the crop from his 150-acre vineyard to produce 36,000 gallons of wine annually. His vineyard, which covers the rolling slope

from the winery building down to the shores of Seneca Lake, is planted to the grape varieties Chardonnay, Riesling, Gewürztraminer, Seyval Blanc, Aurora, Cayuga White, Delaware, Rougeon, de Chaunac, Vidal, and Ravat. From these varieties, Wagner produces both varietal and proprietary still table wines, including Seyval Blanc, Johannisberg Riesling, Capital White, and Wagner's Red, along with a small quantity of sparkling and fortified wines. All wines are estate bottled.

Tours of Wagner's winery begin on the pressing deck for a look at the Willmes press. (Wagner was the first winery in the East to use this advanced low-pressure tank press to extract juice from the grapes.) Next you are shown the tank room, with its large refrigerated stainless-steel tanks used for fermentation. Then the bottling operation is discussed, followed by a tour of the wine cellars, where the wine is aged. The walls of the cellars are lined with oak cooperage, which range in size from 1,500-gallon American oak casks to 135-gallon Yugoslavian oak barrels. Tours end in the tasting room, where between ten and twelve wines, including Wagner's Riesling and Gewürztraminer, are available for sampling. The winery's product list is particularly distinguished by its Chardonnay, Seyval Blanc, and Johannisberg Riesling (all award-winners).

A convenient café, sheltered by a brightly striped tent, is located on a deck beside the winery. Open from June through October, it affords a wonderful view of the vineyards and the lake beyond. There are also picnic facilities on the grounds.

Open all year, except Thanksgiving and Christmas. April 1-December 31: Monday-Saturday 10-4, Sunday 10-5; otherwise Saturdays only 10–4. Guided tours. Free tasting. Retail outlet: wine, gifts. Restaurant. Picnic facilities. Access for the handicapped. Directions: New York Thruway to exit 41. Route 414 south past Lodi to winery. (Wagner is 14 miles north of Watkins Glen on the east side of Seneca Lake.)

Naples

Widmer's Wine Cellars, Inc.
West Avenue
Naples, New York 14512
Telephone: (716) 374-6311
President: William Wickham V

Widmer's Wine Cellars is one of New York State's oldest and largest wineries; it was founded by Swiss-born John Jacob Widmer, who settled in Naples in 1882. Over a century later, the winery's capacity has been estimated at 3 million gallons, but since its present owners consider this figure priviledged information, the estimate cannot be confirmed. Nonetheless, Widmer's facility is obviously substantial. Its modern complex sprawls across the Naples Valley and is surrounded by a portion of its 225-acre vineyards. Widmer also has the distinction of being the only winery in the Naples Valley, its main tourist attraction, and its biggest industry.

Wine tourers will enjoy stopping at Widmer's. It has some unusual features and offers one of the most detailed tours in the Finger Lakes region. Guided tours begin in the winery's reception area, where, while you wait for one of the frequently scheduled tours to leave, you can watch a slide presentation on winemaking or peruse the wine artifacts on display.

The tour, which takes at least an hour, covers each step in the winemaking process, from the vineyard itself, to fermentation, aging, filtration, and bottling. Widmer's famous "cellar on the roof" is included. Visible from the road as you approach the winery, this cellar, used for aging sherries, is actually located outdoors on a section of the winery's roof. Here thousands of sherry-filled oak barrels, stacked four deep, are exposed to the environment for four years in order to "breathe" in the open air before they are blended. The tour is followed by a short wine appreciation course in Widmer's chalet-style tasting room. Three wines are tasted and discussed, and cheese and crackers are served.

The whole line of Widmer wines can be sampled in the adjoining retail shop. Widmer's product line includes such generic wines as Chablis Blanc, Rhine, and Sauterne; proprietary wines (Lake Niagara and Naples Valley Rosé); Charmat process champagne; dessert wines, sherry and port; and vintage-dated varietals (Cayuga White, Johannisberg Riesling, Maréchal Foch). Widmer's most popular product is its proprietary wine, Lake Niagara; it is also noted for its award-winning dessert wines.

Widmer's grows viniferas, Labruscas, and French-American hybrids in its vineyards; additional grapes needed for production are purchased. While picnicking is not permitted on the grounds, there is a public park within walking distance.

Open all year: Monday-Saturday, 10-4; Sunday, 11:30-4:30. Guided tours: Memorial Day-October 31, every half hour or more frequently if required. Free tasting with or without tour. Retail outlet: wine and wine-related gifts. Directions: New York Thruway to exit 46. Take Route 390 south to Wayland exit. Follow signs to Naples and the winery.

Starkey

Hermann J. Wiemer Vineyard
Route 14
Starkey, New York 14837
Telephone: (607) 243-7971
Owner: Hermann J. Wiemer

Hermann J. Wiemer Vineyard is a few miles north of Glenora Wine Cellars on the east side of Lake Seneca. The winery, located in a sixty-year-old barn, has a new dramatic interior, designed by a crew of award-winning Cornell architects to suit its current function. Incorporated in the lofty space are the production facility, offices, laboratory, retail sales and tasting room, and champagne-aging areas.

Hermann Wiemer was born into the winemaking tradition; he was raised in the heart of Germany's Mosel Valley, where his family had been involved in the wine industry for over 300 years. After emigrating to the United States in the late '60s, he served as head winemaker at Hammonds-port's Bully Hill for more than a decade, winning acclaim for the Seyval Blancs and Chancellors he produced. In 1973, Wiemer purchased 140 acres of land along Seneca Lake and planted one of the first all-vinifera vineyards in the area; he left Bully Hill in 1980 to devote full time to his vineyard, expanding nursery, and small winery.

Today, Hermann Wiemer's 7,500-case winery produces only estate-bottled vintage-dated wines made entirely from vinifera grapes, most of which he grows in the surrounding 60-acre vineyard. At present the winery produces award-winning Johannisberg Rieslings (the famous wines of the Mosel Valley), an oak-aged Chardonnay, and a small quantity of *methode champenoise* sparkling wine.

From May through October, informal guided tours of the winery can be had on request between 11 and 5. But even without a guide, many of the working areas of the barn are readily seen from its central tasting/retail area.

Open daily May-October: Monday-Saturday, 11-4:30; Sundays, 12-5. Guided tours on request. Tasting $1 (refunded with purchase of wine). Retail outlet: wine. Picnic facilities. Access for the handicapped. Directions: New York State Thruway to exit 42. Take Route 14 south 20 miles to winery on west side of lake. (Mailing address: Box 4, Dundee, New York 14837.)

LONG ISLAND

Bridgehampton

Bridgehampton Winery
Sag Harbor Turnpike (P.O. Box 979)
Bridgehampton, New York 11932
Telephone: (516) 537-3155
Owner: Lyle Greenfield

Long Island, best known for its potato and cauliflower crops, has recently emerged as an important grape-growing region within New York State. The fashionable Hamptons area on the island's South Fork has been designated by the U.S. Department of Agriculture as an official viticultural region. This appellation allows wineries within the area to label wines made from grapes grown on their estates or other estate-owned land as "estate-bottled."

The owner of Bridgehampton Winery, New York advertising executive Lyle Greenfield, has been a leader in gaining this important designation. He has also given careful thought to the design of the winery. At first glance, it appears to be a modest structure, but closer examination reveals a sophisticated architectural concept. An underground level contains both storage and production facilities; the main floor, a spacious tasting room and retail outlet. The entire rear wall of this level consists of a bank of sliding glass doors opening onto a deck furnished with tables and chairs for visitors. Here you can enjoy views of Bridgehampton's twenty-acre vineyard, a pond, and the woods beyond.

The tour of the winery, which takes about thirty minutes, includes a visit to the vineyard, an inspection of the underground winemaking facilities, and, finally, a sampling of Bridgehampton's product in the main-floor tasting room.

BRIDGEHAMPTON

THE HAMPTONS · 1983 · LONG ISLAND

CHARDONNAY

The Bridgehampton vineyard is planted to the grape varieties Chardonnay, Johannisberg Riesling, Cabernet Franc, Cabernet Sauvignon, Sauvignon Blanc, and Merlot. The winery currently supplements the harvest of its still-maturing vines with grapes purchased from other Long Island growers. In 1984, it released 3,000 cases of wine, all varietal table wines that are vintage dated and estate bottled and produced. Bridgehampton is particularly noted for its Chardonnay, a three-time award-winner, as well as for an excellent Riesling.

Open Spring-December: Monday-Saturday, 11-5. Guided tours, summer and fall. Free tasting. Retail outlet: wine, gifts. Picnic facilities. Festivals: Strawberry-Riesling (late June), Chardonnay (July), Harvest (October), and Nouveau (November). Check with winery for exact dates. Access for the handicapped. Directions: From New York City take the Long Island Expressway to exit 70. South on Manorville Road to Route 27 east. Follow 27 east to Bridgehampton. Winery 1 mile north of Bridgehampton monument.

Cutchogue

Hargrave Vineyard
Alvah's Lane (P.O. Box 927)
Cutchogue, New York 11935
Telephone: (516) 734-5111, 734-5158
Owners: Alec and Louisa Hargrave

Alec and Louisa Hargrave were among the first to recognize Long Island's favorable viticultural conditions—a growing season that averages 210 days (the same as Bordeaux), loamy topsoil, and a temperate climate moderated by the surrounding ocean breezes. In 1973, when the Hargraves purchased a 300-acre potato farm on the North Fork and began planting its acreage to grapes, they were considered pioneers in the area. But since then more than twenty growers have followed their lead, and two more wineries have been bonded.

The Hargraves' vineyard and winery are located near the small town of Cutchogue; the forty-five acre all-vinifera vineyard is planted primarily to Chardonnay and Cabernet Sauvignon. And the former potato barn, with its underground cellar, has been adapted comfortably to its new life as a winery.

Hargrave's first wine, a Pinot Noir, was bottled in 1975; the product line now includes Riesling, Sauvignon Blanc, Fumé Blanc, Cabernet Sauvignon, and Merlot. The winery produces 7,000 cases of wine annually; its wines consistently win awards and are available in many fine New York restaurants. The Collector Series Chardonnay is particularly noteworthy.

One tour of the winery is given each weekday afternoon; three are offered on Saturday and Sunday. These take approximately ninety minutes and begin with a visit to the vineyard. The winemaking operation is explained in detail; Louisa Hargrave is frequently on hand in the tasting room to comment on the wines as you sample them.

Hargrave Winery has picnic facilities and in summer months hosts a series of chamber music concerts in the vineyards.

Open May-November: daily 10-5. Guided tours: weekdays at 2; weekends at 12, 2, and 4. Free tasting. Retail outlet: wine. Chamber music concerts (summer). Picnic facilities. Access for the handicapped. Directions: Take Long Island Expressway east to exit 71 (Edwards Road); go north across Route 25 to Sound Avenue. Turn right onto Sound Avenue (going east) and continue for 12 miles. (Sound Avenue will become County Road 48, a four-lane highway.) Proceed for 3 miles; the vineyard will be on the right side of the highway.

Peconic

Lenz Vineyards
Main Road
Peconic, New York 11958
Telephone: (516) 734-6010
Owners: Patricia and Peter Lenz

Patricia and Peter Lenz, founders of Lenz Vineyards, had been the owners/chefs of a popular Westhampton restaurant specializing in regional food. Their interest in and enthusiasm for American cuisine led them to begin their own winemaking venture in 1977. The couple purchased a farmhouse and outbuildings on a thirty-acre potato farm near Cutchogue on Long Island's North Fork. In conjunction with architect Mark Simon, they renovated and refitted the outbuildings to suit the new working winery. The resulting architecture, ingeniously connected by trellising, has won several design awards, and makes a stop at Lenz Vineyards even more pleasant.

Lenz produces 15,000 gallons of table wine annually from vineyards planted to the European grape varieties Gewürztraminer, Pinot Noir, Chardonnay, Merlot, Cabernet Sauvignon, and Cabernet Franc. Its Gewürztraminer, Merlot, and Lenz Reserve are especially noteworthy.

Lenz offers only informal tours (tastings are by appointment); plan to stop here briefly before or after visiting Hargrave Vineyard in neighboring Cutchogue.

Open April-November: daily 11-4. Informal tours (summer months only; reservations suggested). Retail outlet: wine. Directions: On Route 25, 1½ miles east of Cutchogue.

Other Middle Atlantic States

Each of the Middle Atlantic states of New Jersey, Pennsylvania, and Maryland has developed differently as a winegrowing region despite similar climatic conditions. Pennsylvania, the largest and most important of the three, ranks fifth in the nation in grape cultivation. It is the state where American commerical grapegrowing began, these efforts dating back to 1683 when William Penn brought French and Spanish vines to the Philadelphia area. But until 1963, Pennsylvania did not have even *one* producing winery. It was, and still is, largely a grape juice, jelly, and jam state. And its greatest concentration of vineyards, in the far northwestern corner of the state around the town of Northeast, is part of the grape belt that extends along the Lake Erie shore from New York to Ohio, an area with the largest plantings of Concords in the world.

The repeal of Prohibition did not significantly change the winegrowing picture in Pennsylvania since the legislature restricted the sale of wine and alcoholic beverages to state-owned stores. Not until the passage of the Limited Winery Act in 1968 did the state's grapegrowing profile began to shift slightly. Under the new regulations, a farm winery was permitted to sell its own products at the facility. A subsequent amendment passed in 1982 allows a winery to open as many as three retail/tasting room outlets in other locations.

Today Pennsylvania has 9,000 acres planted to vineyards and thirty-one bonded wineries. Roughly eighty-two percent of the grapes grown continue to be Concords, but the new vineyards are largely planted to wine grapes, including French-American hybrids, vinifera, and less foxy native American varieties such as Catawba and Delaware. As one might expect, some of these new wineries have been established in the proven grapegrowing area of Northeast, Presque Isle and Penn Shore being two of the best-known new facilities. But the southeastern corner of the state—in particular agriculturally rich Lancaster County—is emerging as a fertile region for winegrowing. The wineries there are among Pennsylvania's most picturesque and hospitable and include Mount Hope, a large landmark estate with handsome, formal gardens; small Tuquan, on a family farm; and Nissley, a 300-acre cattle farm whose winery is tucked into an 18th-century tobacco barn.

New Jersey's wine industry dates back to 1855 and the founding of the Renault Winery by Frenchman Louis Nicholas Renault in the town of Little Egg Harbor. This historic winery even managed to survive the dry years of Prohibition by producing a popular nerve tonic sold in drugstores. Unlike Pennsylvania, New Jersey did not develop a grapegrowing industry, although Vineland in Cumberland County is the place where Dr. Thomas B. Welch established his fresh grape juice business. A grape rot in the late 1880s forced him to depart for the more promising grapegrowing region of New York's Finger Lakes.

The winemaking revolution which swept many areas of America in the 1960s reached New Jersey in 1981 with the passage of a state winery bill. This legislation has given impetus to the founding of new wineries located in the northwestern counties of Hunterdon and Warren. Here historic dairy and horse barns have been converted to winemaking facilities, and former pasture land on the sheltering slopes of the Musconetcong Valley and along the Delaware River is proving suitable for the planting of French-American hybrids and even vinifera. These new wineries, including Tewksbury (featured on the cover of this book), Alba, and King's Road are smaller than their counterparts in South Jersey, but they are rewarding to visit for their rural character, handsome buildings, scenic settings, and award-winning wines.

The small state of Maryland has only ten wineries and approximately 350 acres planted to grapes, but it has a rich and important wine history. A Marylander, Major John Adulm, discovered and named the famous American grape, Catawba. He was also the author of the first American book on winegrowing, *A Memoir on the Cultivation of the Vine and the Best Mode of Making Wine* (1823).

Today there are two viticultural regions in the state—Catoctin and Lingamore—and its vineyards are planted primarily to French-

American hybrids. Philip Wagner, founder of the Boordy Vineyards, led in the introduction of these varieties, and cuttings from his plantings provided the rootstock for many other Eastern vineyards established after Prohibition. All of the Maryland wineries are within easy reach of Baltimore and Washington, and all are worth visiting.

MARYLAND

Brookeville

Catoctin Vineyards
805 Greenbridge Road
Brookeville, Maryland 20729
Telephone: (301) 774-2310
Owners: Ann and Jerry Milne, Judy and Roger Wolf

One of Maryland's newest wineries with high-tech production equipment. First production in 1983: 7,500 gallons. Vineyard (twenty-five acres) remote from winery in Catoctin viticultural district, planted to Chardonnay, Cabernet Sauvignon, Johannisberg Riesling. Winery in Swiss chalet-style building formerly owned by Provenza Winery. Winemaker, Bob Lyon, formerly at Chateau Montelena in the Napa Valley. Winery noted for oak-fermented Chardonnay. Open all year: daily 10-6. Short guided tours and tastings ($2 charge for wineglass) regularly scheduled: Saturday, Sunday, and holidays; other days by appointment. Retail outlet: wine, gifts. Picnic facilities. Access for handicapped. Directions: From Baltimore Beltway (Interstate 695) take Interstate 70 to Route 29. Follow Route 29 south to Route 108 (towards Columbia). Follow 108 west to Route 650. Take 650 (also called New Hampshire Avenue) north 4 miles to winery signs.

Hydes

Boordy Vineyards
12820 Long Green Pike
Hydes, Maryland 21082

Telephone: (301) 592-5051
Owner: Robert F. Deford III

Boordy, Maryland's oldest and largest winery, is also one of the most important in the East. It was founded by Philip Wagner, a former editor on the *Baltimore Sun*, who is largely credited with the introduction of the French-American hybrid grapes now grown extensively in this part of the country.

Wagner loved fine wine, and in the 1920s Prohibition forced him to try his hand at home winemaking. After Repeal, he attempted to grow California vinifera varieties in Maryland, but was soon thwarted by the climate. He then turned successfully to growing French-American hybrids imported from Bordeaux. By 1946, he and his wife, Jocelyn, had established a hybrid nursery, bonded a winery next to their home in Riderwood, and had begun selling their wines to Baltimore restaurants. A taste for their vintage wines soon spread to Washington and New York. The business of supplying eastern winegrowers with hybrid stock, however, grew even faster than the winery. In 1964, Wagner retired from the *Baltimore Sun* to devote himself entirely to his winery, nursery, and consulting business for fellow winemakers. Along the way he also wrote *American Wines and How to Make Them*.

In 1980 the Wagners sold everything but the nursery to one of their long-time growers, the Deford family, who promptly sent their son, Robert Deford III, to the University of California at Davis for training in winemaking. The winery was relocated from Riderwood to the Deford Vineyard in nearby Hydes, and here you will find the fine Boordy tradition still being carried on.

Boordy is located in the rolling hills of Baltimore County. The winery is a handsome renovated 19th-century barn with thick stone walls and heavy beams and is surrounded by twelve acres of vineyard. The acreage is planted to Cabernet Sauvignon, Riesling, Seyval Blanc, Vidal Blanc, Chancellor, Chambourcin, and Maréchal Foch. Tours of the premises, which usually take upwards of forty-five minutes, begin in the vineyard (weather permitting) with a history of the grapes planted, their propagation, cultivation, harvest, and specific uses.

Within the winery itself, every step of the winemaking process is covered from crushing the fruit to bottling and cellaring the finished wines. This is followed by a sampling of several of Boordy's current wines, and includes, when possible, barrel samplings of young wines. Boordy produces 15,000 gallons of still table wine annually and is particularly known for its Nouveau, Vin Gris, and the varietal Vidal Blanc.

Visitors are welcome to picnic on the grounds. Various annual festivities (admission by reservation only) include a Nouveau Celebration (fall), French Country Dinners (winter), and Open House (spring). Write or call for calendar of events.

Open all year: Tuesday-Saturday, 10-5; Sunday, 1-4; Guided tours. Free tasting. Retail outlet: wine, gifts, books. Special events by reservation. Picnic facilities. Access for the handicapped. Directions: Baltimore Beltway (Interstate 695) to exit 29 (Cromwell Bridge Road). East on Cromwell Bridge Road 2.9 miles to end. Left on Glen Arm Road to Long Green Pike. Left (north) on Long Green Pike 2 miles to winery.

Westminster

Montbray Wine Cellars
818 Silver Run Valley Road
Westminster, Maryland 21157
Telephone: (301) 346-7878
Owners: G. Hamilton and Phyllis N. Mowbray

Bonded in 1966, and housed in a classic Pennsylvania-German barn, Montbray Wine Cellars is located in the hills of the Silver Run Valley two miles from the Pennsylvania border and not far from the Gettysburg historic sites. The barn is built into the bank of a hill, and its underground level houses the winery's working facility. Montbray's thirty-acre vineyard is planted to Cabernet Sauvignon, Chardonnay, Riesling, Cabernet Franc, and Seyvre-Villard. From this crop, and additional purchased grapes, is produced vintage-dated varietal tables wines. The winery is especially noted for its Seyvre-Villard (Seyval Blanc) and Cabernet Sauvignon.

Tours of Montbray are quite informal and personalized and are usually guided by one of the Montbrays or their cellarmaster, who tailor their talks about winemaking and the winery to the special interests of the visitors. Afterwards, from three to six wines are sampled, depending on what is currently available. Picnicking is permitted in the walnut grove next to the winery.

Open all year: Monday-Saturday, 10-6; Sunday, 1-6. Guided tours (appointment suggested one week in advance). Free tasting. Retail outlet: wine and wine-related items. Directions: From Westminster, take Route 97 north 7 miles to Silver Run Valley Road. Turn right on Silver Run Valley Road; 2 miles to winery (on your left).

NEW JERSEY

Absecon

Bernard D'arcy Wine Cellars
306 E. Jim Leeds Road
Absecon, New Jersey 08201
Telephone: (609) 652-1187
Owner: Bernard F. D'arcy

On the outskirts of Atlantic City, not far from the Renault Winery in Egg Harbor City, is Bernard D'arcy Wine Cellars. (A second sales outlet is located in Manasquan.) Founded in 1934 by the late Johann Gross, the winery is now owned and operated by Gross's grandson, Bernard"Skip" D'arcy. The firm is the largest direct-to-the-consumer winery in the East. The proximity of the winery to Route 9 and the Garden State Parkway, well-traveled arteries to Atlantic City, undoubtedly accounts in part for its impressive sales record. A more important contributing factor to the company's success is its comprehensive selection of over twenty wines from white and red to dessert and sparkling, many of which have received awards.

The visitor to Bernard D'arcy Wine Cellars in Absecon will discover a cluster of picturesque buildings surrounded by mature shade trees. A row of German ovals lined up outside, though empty, dispels any lingering doubts about what activities are conducted in the nearby structures. Tours of the facility are self-guided and include the bottling line (upstairs) and the barrel aging cellar (downstairs). While there are no educational signposts to guide your way through these public areas, there is a short slide presentation afterwards that explains the winemaking process. The fermentation tanks are located in another building not open to the public.

Bernard D'arcy's whole line is available for sampling in the tasting room. You serve yourself from open bottles of still table wines on the bar, a most generous allowance by the winery. Sparkling wines are available, too, but these are poured on request so that they don't lose their effervescence by standing open for long periods.

In the adjoining retail area Bernard D'arcy's wines can be purchased in single bottles, case lots, and gift packs. Among the award-winning wines are Brut Champagne, Rhine Wine, Niagara, and Cream Almonique (a dessert wine). A separate gift shop is located in front of the main winery building and carries a wide selection of glassware, cookbooks, and wine-related items.

Open all year: Monday-Saturday, 10-6. Self-guided tours. Slide presentation.
Free tasting. Retail outlets: tasting room, wine only; Glass House, wine-related
merchandise, souvenirs. Access for the handicapped. Directions: From Garden
State Parkway, southbound to exit 40. U.S. 30 East to 6th Avenue. Left on
6th Avenue to winery. Manasquan outlet (tasting and sales only): Route 35
north of Brielle Circle, Manasquan. Open all year: Monday-Saturday, 10-6.

Egg Harbor City

Renault Winery
Bremen Avenue (R.D. 3, Box 21B)
Egg Harbor City, New Jersey 08215
Telephone: (609) 965-2111
Owner: The Joseph Milza Family

Roots reaching back to the mid-1800s make Renault Winery in Egg Harbor City the oldest winery in continuous operation in New Jersey and one of the oldest in the nation. (Renault was the third winery bonded in the United States.) A stone's throw from the Atlantic City boardwalk and casinos and some of New Jersey's finest beaches, the winery attracts many visitors each year with its very hospitable features. In addition to tours and tastings, Renault hosts a series of festivals throughout the year, operates three restaurants at the winery, and displays a fascinating collection of wine and champagne glasses, some dating to the 13th century, in a museum on the premises.

Renault's 120-year history began when Louis Nicholas Renault, representing the ancient champagne house of the Duke of Montebello at Rheims, France, came to the United States to establish a vineyard free of phylloxera (the aphid which was ravishing the vineyards of Western Europe). After an unsuccessful attempt at growing disease-free grapes in California, Renault came East and decided to start his own vineyard in New Jersey planted to the hardy native American Labrusca grape. Renault's vineyard thrived. His wines won many awards, and he became the largest distributor of champagne in the United States. (In his day, Egg Harbor was known as "the wine city.")

Louis Renault died in 1913 at the age of ninety and was succeeded by his son Felix who sold the company six years later to John D'Agostino. Throughout Prohibition a government permit allowed the winery to produce "Renault Wine Tonic." Sold in drugstores, it had an alcoholic content of twenty-two percent and was so popular that the winery had to operate twenty-four hours a day to fill the demand for this "health elixir."

Currently Renault is owned by Joseph P. Milza, a former newspaper owner and publisher. The winery is located on a 1,000-acre farm, of which 130 acres are planted to Noah, Baco Noir, de Chaunac, Catawba, and Niagara-Norton. The winery produces over 50,000 gallons of wine annually, fifty percent of which is still table wine and fifty percent sparkling wine.

Guided tours of the winery, which cost a dollar, begin with the antique glass museum in which the collection of champagne and wine glasses is displayed. It then progresses to the hospitality room, with its large German ovals, where Renault's history is recounted. Continuing on through the winery's atrium, it features yet another impressive display, this one consisting of old-world winemaking equipment. Included are antique wine presses, dosage machines, and gravity filters. After an explanation of how wine was made with the antique apparatus, the tour moves into the pressing room for a look at 20th-century practices. The tour concludes with a glimpse of Renault's huge wine cellar and a visit to the tasting room for samples of the firm's product.

Renault's product line includes the varietal wines Noah and Catawba, and the generics Chablis, Burgundy, and Rose. The firm is particularly known for its White Champagne, Blueberry Champagne, and Noah.

In addition to the working winery, exhibit rooms, and sales area, there are three restaurants on the site—the Methode Champenoise, a dinner restaurant located on the top floor of the winery; the Garden Cafe, which is open for lunch; and the Wine Cellar, which is used for catered parties and the winery's Saturday night candlelight tours and buffets. The Methode Champenoise is the most elaborate of the three and offers an extensive wine list and *prix fixe* dinners including an appetizer, pasta, salad, soup, sorbet, entrée, and dessert.

Open all year, except Thanksgiving, Christmas, New Year's and Easter: Monday-Saturday, 10-5; Sunday, 12-4. Guided tours daily $1 for adults, children free); last tour at 4. Free tasting. Retail outlet: wine, gifts. Picnic facilities. Three restaurants: Methode Champenoise, Garden Cafe, and the Wine Cellar. Reservations are suggested for the Methode Champenoise room and the Wine Cellar buffet. Festivals: numerous throughout the year. Write for calendar. Access for the handicapped. Directions: Garden State Parkway south to exit 44 (Egg Harbor City). Follow Moss Hill Road (a sharp right turn off the Parkway) 5 miles to Bremen Avenue. Turn right on Bremen Avenue. Winery 2¼ miles on your right.

Finesville

Alba Vineyard
Route 627
Finesville, New Jersey 08848
Telephone: (201) 995-7800
Proprietor: Rudolf Marchesi

Alba Vineyard, located on an eighty-acre tract in the hills of western New Jersey, is a new winery in an old form. The winery's handsome century-old stone and wood buildings situated in the Musconetcong Valley near the meandering river of the same name, once sheltered cows and farm equipment. Carefully restored and modified for their new use, the deceptively rustic buildings now comprise a thoroughly modern working winery, including a spacious combination tasting room/art gallery/sales area.

Beyond the winery on a southern slope where cows once grazed is Alba's thirty-five acre vineyard. Planted here are viniferas and French hybrids, namely Chardonnay, Vidal Blanc, Gewürztraminer, Cabernet Sauvignon, Merlot, and Foch. The young vineyard currently provides only a portion of the grapes needed for the winery's annual output of 100,000 bottles. The remaining grapes necessary for production are purchased.

Alba, bonded in September, 1983, was founded by Rudolf Marchesi, who began planting his vineyard in 1979. Marchesi grew up in Mahwah, New Jersey, "watching my grandparents, who came from a family of winemakers and growers in Lombardy, Italy, making wine." This early exposure to winemaking left its mark. After majoring in psychology in college, Marchesi planted a small experimental vineyard (1½ acres) in California, but drought caused his well to dry up and curtailed this agricultural endeavor. And he then came back East with his wife, Wendy, a certified nurse-midwife, and started a winery in his home state.

A tour of the Albas' well-planned facility is particularly educational. The visitor is given an in-depth look at the working winery with its cut-stone aging cellar filled with oak barrels, the stainless-steel fermentation tanks in an adjoining room, and the bottling line. The complete tour of the winery and vineyards takes about 45 minutes. For the children there are even hayrides through the vineyards. Afterwards, Alba's product can be sampled at its stand-up bar which also houses a changing exhibition of local artists' work. Picnic facilities include tables and benches under an arbor behind the winery and amid the vineyards on a grassy knoll overlooking the Delaware.

Alba's wines include a Proprietor's White and Red; Proprietor's Reserve White and Red; the varietals Seyval Blanc, Chardonnay, and Riesling; various fruit wines; and a Spring Red and a Rosé.

Open all year: Wednesday-Sunday, 12-6. Guided tours. Free tasting. Retail outlet: wine, gifts. Art gallery. Picnic facilities. Art/jazz/wine weekend at summer's end. Access for the handicapped. Directions: From Interstate 78, exit 7 (Bloomsbury). Route 173 west for 1.3 miles to Route 639. Left on Route 639 west to stop sign (junction of Routes 519 and 627 South). Follow 627 South for 2.4 miles (winery on your right). Mailing address: R.D. 1, Box 179AAA, Milford, New Jersey 08848.

Hammonton

Tomasello Winery
225 N. White Horse Pike
Hammonton, New Jersey 08037
Telephone: (609) 561-0567
Owners: The Tomasello Family

Tomasello Winery in Hammonton was established in 1933 by Frank Tomasello, a southern New Jersey vegetable farmer with Sicilian roots. True to his Italian heritage, Tomasello first made wine from a small planting of grapes on his property. Although fermented only for his personal consumption, the wine was much admired by his friends. Winemaking soon became more than a sideline, and before long grapes replaced his other agricultural crops. With the repeal of Prohibition, Tomasello decided to go public with his product.

The winery has grown under the management of his sons Charles and Joseph, and Charles's children, Charles Jr., Jane D., and John K. Tomasello. The operation has also taken on a more scientific tenor. French hybrids have been planted in the vineyards, *methode champenoise* sparkling wines introduced, and the product line diversified.

At its Hammonton location (the winery has an additional outlet for tasting and sales in historic Smithville) 100 acres of grapes are under cultivation from which the winery produces 80,000 gallons of wine annually. Varieties grown are primarily both American and French hybrids with a sprinkling of vinifera. From these Tomasello produces six *methode champenoise* sparkling wines and twelve still table wines—including two vintage-dated varietal wines, Villard Blanc 1982 and de Chaunac 1982, and such generic wines as Rhine Chablis and Burgundy.

Guided tours of the working winery are by appointment only and cover the whole facility from grape processing to bottling. Free tasting and sales are available anytime the winery is open. More tourist-oriented is Tomasello's recently opened outlet in Smithville, north of Atlantic City. Free tastings of the winery's entire line can be had here, too, and there is a mini-museum featuring old-fashioned winemaking equipment and a pictorial exhibit explaining modern wine production. The outlet also hosts numerous festivals.

Hammonton: *Open all year, Monday-Saturday, 9-8; Sunday, 12-6. Guided tours by appointment only (24 hours notice suggested). Free tasting. Retail outlet: wine. Directions: From northern New Jersey: Route 206 south to Route 30. West on Route 30. Winery on your right, ½ mile past marker #30.* Smithville: *Open all year, Winter, 12-6; Summer, 11-9. Free tasting. Retail outlet: wine. Museum. Festivals. Access for the handicapped. Directions: Garden State Parkway south to Exit 48 (Smithville). Follow Route 9 and the signs to Historic Towne of Smithville.*

Lebanon

Tewksbury Wine Cellars
Burrell Road (R.D. 2)
Lebanon, New Jersey 08833
Telephone: (201) 832-2400
Owner: Daniel F. Vernon, Jr., DVM

An idyllic pastoral scene, composed of 18th-century barns, a stone farmhouse, and a horse paddock and pasture complete with pond and splashing ducks, is home to New Jersey's award-winning Tewksbury Wine Cellars. Although its mailing address in the rolling hills of Hunterdon County is Lebanon, it is actually closer to the village of Oldwick.

Founded in 1979 by Dan Vernon, a veterinarian, the winery's twenty-acre all-vinifera vineyard is planted to the European grape varieties Chardonnay, White Riesling, Gewürztraminer, Rayon d'Or, Chambourcin, and Pinot Noir. From these grapes—which are occasionally supplemented with some purchased fruit—Dr. Vernon produces 80,000 gallons of wine annually. Nearly eighty percent of his product line is devoted to estate-bottled varietal table wines, but a few fruit wines round out the list.

Tours of the facility include the winery—and the vineyards, if guests are so inclined—and cover the vinification process from start to finish. The winery, located in a barn behind the tasting/retail room, once served as a horse hospital. In the former operating theatre, medical trappings have been replaced by fermentation tanks. The horse recovery stalls around the room's perimeter are now used for case storage. In an adjoining barn, additional recovery stalls now serve as the barrel-aging cellar. The wines are generally aged in new, smaller oak barrels which impart more flavor than the large old-fashioned German ovals which are also stored here.

After the tour you can sample Tewksbury's product. (There is a $2 charge per person for tasting, but this is refunded with the purchase of two bottles of wine.) The winery's product list includes numerous award-winning wines. The 1983 Chardonnay received a Gold Medal and Best of Class Award in a recent international wine competition. Other recent winners have been a 1982 Chambourcin and a 1981 Gewürztraminer. Other premium table wines include Chenin Blanc, White Riesling, Delaware, Rayon d'Or, Rosé de Chardonnay. Tewksbury also produces three fruit wines: Oldwick Apple Wine (made from apples from a local orchard which was bearing fruit before George Washington's time), Peach Wine, and Cranberry Apple Wine.

After touring and tasting you are free to picnic on the lawn or by the pond and to pet the friendly resident cows and horses and feed the ducks. Tewksbury Wine Cellars also hosts a number of festivals during the summer months such as the Grape Blossom Festival (June) and Sangria Weekend (August). Call or write for a current calendar of festival dates.

Open all year: Wednesday-Friday (call ahead for information on hours); Saturday, 11-5; Sunday, 1-5. Guided tours. Tasting. Retail outlet: wine, gifts. Picnic facilities. Directions: Interstate 78 to Exit 24. North on Route 517 through village of Oldwick to Sawmill Road. Left on Sawmill ½ mile to Burrell Road. Right on Burrell .6 mile to winery (on your right).

Milford

King's Road Vineyard
Route 579 (R.D. 2, Box 352B)
Milford, New Jersey 08848
Telephone: (201) 479-6611
Owners: John and Marie Abplanalp

King's Road, New Jersey's newest winery, was bonded in December, 1984. It is located a few miles northwest of Alba Vineyard, and, like its vinous neighbor, also makes its home in a century-old dairy barn on the sheltering slopes of the Musconetcong Valley. The barn, built into the bank of a hill, is an exceptional example of New Jersey farm architecture. Although recently painted a very stylish taupe, its structural elements—huge hand-hewn beams, shake roof, and fieldstone foundation—attest to its true age and rural roots.

The winery and its surrounding vineyard put to new use a long-neglected property that has been in the family for generations. German-born Klaus Schreiber was chosen by John Abplanalp's father to turn the property into a winery. After intensive courses in viticulture and vinification at Cornell University's School of Agriculture, Schreiber, a mechanical engineer, supervised the planting of the vineyard and the transformation of the one-time dairy barn into a winery. He is also the winemaker.

Three acres of King's Road's vineyards are planted primarily to the grape varieties Aurora, Seyval Blanc, Niagara, and Villard Blanc, with smaller, experimental plantings of Chardonnay, Riesling, and Sauvignon Blanc. To date, the only wines to be released are white. These include Aurora, Seyval Blanc, Niagara, Villard Blanc, and Chardonnay. Tours of the compact winery, located in the depths of the barn, reveal very modern equipment including some very unusual space-saving rectilinear fermentation tanks; a newly excavated underground aging cellar; and a pocket-sized laboratory.

The name of the winery refers to a road that has threaded the property for hundreds of years. Originally an Indian path, King's Road became an important artery during the American Revolution when it was used to haul pig-iron from the furnace at Bloomsbury to the forge at Pittstown.

Open all year, except major holidays: Saturday and Sunday, 12-5. Guided tour. Free tasting. Retail outlet: wine. Limited access for the handicapped. Directions: Take Interstate 78 to Pattenberg, exit 11. Follow Route 614 south to junction of Route 614 and 579 (approximately 3 miles). Go west on 579 for .2 mile. Winery will be on your left.

PENNSYLVANIA

Bainbridge

Nissley Vineyards and Winery Estate
Maytown-Bainbridge Road (R.D. 1, Box 92B)
Bainbridge, Pennsylvania 17502
Telephone: (717) 426-3514
Owners: The Nissley Family

Nissley Vineyards is nestled in a small valley on a bucolic 300-acre cattle farm eight miles north of Columbia in Pennsylvania-Dutch Country. The winery was founded nearly thirteen years ago by J. Richard Nissley—the retired head of the Bear Creek Construction Company, a local bridge-building firm—and his son, John. The senior Nissley had always made wine as a hobby, an endeavor to which he now devotes his full time, while John tends the vineyards.

The Nissleys began planting their vineyards in 1972; four years later they converted the Lancaster County 18th-century stone tobacco barn on the property into a winery. Nissley Vineyards is now one of the largest wineries in the state and also one of its most handsome. The restored stone and wood winery sits amid a complex of equally attractive and historic buildings, including a wood-frame corn barn, an 18th-century stone house, a stone mill of the same vintage, and its companion single-lane stone-arch bridge which J.R. Nissley, with his construction skills, painstakingly repaired.

Completing the estate's pastoral setting are fifty-two acres of vineyards planted to Riesling, Chardonnay, Seyval Blanc, Aurora, DeChaunac, Chancellor, Concord, Niagara, and Vidal. From these grapes the Nissleys produce 30,000 gallons of wine annually. Ninety percent of their product line is devoted to table wine, including vintage-dated varietals such as Seyval Blanc, Aurora, DeChaunac, Chancellor, Chardonnay, and Riesling. They also make proprietary wines such as Naughty Marietta and Keystone Kiss and a small quantity of fruit wine. Nissley Vineyards is particularly known for the varietal wines made from its French hybrid grapes; many have won awards.

Guided tours take about forty-five minutes and include a tour of the vineyard and the operating winery. In summer months, wine tasting, which follows the tour, is conducted outdoors under the barn's graceful stone arches. Picnicking is permitted on the well-groomed lawn next to the winery; you are welcome to spread a blanket or sit at one of the picnic tables. Nissley hosts numerous festivals and special events including Saturday evening lawn concerts in July and August and a road run in October.

Open all year: Monday-Saturday, 12-6. Guided tours. Free tasting. Retail outlet: wine, wine-related merchandise. Picnic facilities. Special events: festivals, May and September; lawn concerts, Saturday evenings July and August; road run, October. Access for the handicapped. Directions: From Lancaster Route 30 west to Columbia exit. North on 441, 8 miles to Wickersham Road. Right onto Wickersham and follow signs to the winery.

Chadds Ford

Chaddsford Winery
Route 1 (P.O. Box 229)
Chadds Ford, Pennsylvania 19317
Telephone: (215) 388-6221
Owners: Eric and Lee Miller

Chaddsford Winery, nestled in the historic Brandywine region of Pennsylvania, is situated just outside the tiny village of Chadds Ford on U.S. Route 1. The small boutique vineyard (yearly output 85,000 bottles) has only been open to the public since 1983, but it is "the culmination of two years work and a lifetime of dreaming," according to its owners, Eric and Lee Miller.

The Millers share a rich background in wine knowledge and lore. Eric Miller's interest in wine evolved naturally: as a child, he lived near many of the famous vineyards of Europe. In the early 1970s, when his family returned to this country, he and his parents established New York State's first farm winery, Benmarl Vineyard. His wife, Lee, is a wine journalist and co-founder of the magazine *Wine East,* as well as co-author of the first book about the wineries of the eastern United States, *Wine East of the Rockies.*

The Millers' goal is to produce world-class wines. To that end they chose their southeastern Pennsylvania location with great care, after studying the soil and climate maps of the entire East Coast and visiting the different regions that seemed promising.

Eric Miller designed Chaddsford winery and its equipment to produce the two types of wine that he believes can be grown best in this region: crisp, light, fresh, fruity wines for early consumption; and medium-bodied, earthy, barrel-aged wines which will improve with time. Ninety-five per-

cent of the winery's output is table wine. Recent vintage wines include Chardonnay and Cabernet Sauvignon (both oak-aged), and several light whites such as Seyval Blanc. Also produced is a Country Rouge, a Country White, and an apple wine.

Currently, the winery purchases all of its grapes from other Pennsylvania growers. The Millers hope in the near future to establish their own vineyard. About twenty percent of the cellar space is devoted to "collector" wines; these include the Chardonnay and Cabernet, which are finished in French oak barrels from the Nevers forest, and the Chambourcin Reserve, which is aged in American oak from the Ozarks.

Chaddsford Winery is located in what the Millers describe as a "lovingly restored" old barn. Here visitors tour the cellar, see the production equipment, view a bottling demonstration, and sample the wine. The winery also offers a full schedule of specialty tastings and "New Year's Eve in the Wine Cellar" in addition to its daily free tours and tastings.

Open all year: Tuesday-Friday, 10-5, Saturday and Sunday, 12-5. Guided tours (appointments necessary on weekdays). Free tasting. Charge for special tastings and events for which reservations are also necessary. Retail outlet: wine and wine-related items. Picnicking on grounds with permission. Limited access for the handicapped. Directions: Just outside Chadds Ford on U.S. Route 1, between Longwood Gardens and the Brandywine River Museum.

Holtwood

Tucquan Vineyard
Tucquan Road (R.D. 2, Box 1830)
Holtwood, Pennsylvania 17532
Telephone: (717) 284-2221
Owners: Thomas and Lucinda Hampton

Three miles south of Rawlinsville in Lancaster County is the tiny (5,000 gallon) winery and vineyard of Thomas and Lucinda Hampton. Tom, a millwright by profession, and Lucinda, his wife, founded Tucquan Vineyard in 1968. Four years later, the couple built the barn winery, including a deck for picnicking, behind their farmhouse. A smaller building to the left of the winery serves as the tasting and sales room.

Located on Drytown Road (the name refers to the dry ridge the property sits on, rather than to any teetotaling inhabitants past or present), the winery is surrounded by ten acres of vineyards planted to French hybrids and Labrusca varieties, including Seyval Blanc, Maréchal Foch, Chancellor, Verdelet, Vidal, Steuben, Concord, and Niagara. All of the wine produced is from grapes grown on the Hamptons' estate. Their wines include the varietals Seyval Blanc, Chancellor, Maréchal Foch, Steuben, Catawba, Niagara, and Concord. They also make a small amount of estate-bottled peach wine from the fruit grown in their orchard. Tucquan's product list is distinguished by three gold-medal winners: Chancellor, Seyval Blanc, and Peach.

The Hamptons like to give the tours of their winery themselves—a personal touch you will get only at a small winery—so they ask that you call one day ahead for an appointment. The tour of the farm takes about thirty minutes and covers the winery, vineyard, and tasting room. The open, shaded picnic deck on the second floor affords a good view of the trellised vineyard, the outdoor crush pad, and press; in clear weather you can see the surrounding countryside for thirty miles.

Open all year: Monday-Saturday, 11-5. Guided tours by appointment only (one day's notice). Free tasting (barrel tastings by appointment). Retail outlet: wine, gifts. Picnic deck. Access for the handicapped. Directions: From Lancaster, take Route 222 to Route 372. Left on 372, 4 miles to Hilldale Road. Turn right on Hilldale, then left on Drytown.

Manheim

Mount Hope Estate & Winery
Route 72
Manheim, Pennsylvania 17546
Telephone: (717) 665-7021
President: Charles J. Romito

Mount Hope Estate & Winery, located in the heart of farmland populated largely by the simple Amish sect, will appeal to history buffs and antique lovers as well as to wine connoisseurs, for a visit here includes a tour of an historic mansion and its adjacent formal gardens.

Mount Hope Estate was built in 1800 as the summer home of Henry Bates Grubb, one of 18th-century America's wealthiest ironmasters, on land purchased by Peter Grubb in 1779; Peter Grubb was the founder of Cornwall Furnace, now a state historic park, located nearby. The thirty-two-room sandstone dwelling, listed on the National Register of Historic Places, was occupied by five generations of Grubbs. The last family member to own the estate, Daisy Grubb, Victorianized the Federal mansion, which still contains much of the family furniture. Among the notable architectural features are castle-like walls and turrets, a winding walnut staircase, hand-painted eighteen-foot ceilings, Egyptian marble fireplaces, a grand ballroom, and imported crystal chandeliers. Equally interesting are Mount Hope's gardens, planted with rare and exotic shrubs from all over the world. Included in the twenty-five acre horticultural display encircling the house are specimen plantings of English boxwood, cypress and sassafras trees, and umbrella and candelabra pines.

Mount Hope Estate was purchased in 1980 by Lebanon attorney Charles Romito and a group of local investors. They not only restored the mansion to its former elegance, a major task in itself, but started the winery and planted vineyards. Planted on the perimeter are ten acres of French hybrid grapes—Riesling and Vidal Blanc. These crops are supplemented with fifty acres of grapes purchased from other Pennsylvania growers to yield the 250,000 bottles produced here annually.

Open all year, except Thanksgiving, Christmas, New Year's, and Election Day: Monday-Saturday, 10-4:30; Sunday, 12-4:30. Admission to grounds: adults $2; children (6-14) $1. Guided tours of winery and mansion. Vintage Wine Shop and Mansion Gift Shop: wine, wine-related items, picnic hampers. Picnicking in formal gardens permitted. Many special events, such as Pennsylvania Renaissance Fair weekends in September, and summer concerts. Directions: Pennsylvania Turnpike to Exit 20. South on Route 72, ½ mile to winery. (Mailing address: P.O. Box 685, Cornwall, Pennsylvania 17016.)

New Hope

Bucks Country Vineyards & Winery
Route 202 (R.D. 3, Box 167)
New Hope, Pennsylvania 18938
Telephone: (215) 794-7449
Owner: Arthur Gerold

Bucks Country Vineyards sits beside the heavily trafficked tourist stretch of Route 202 between the popular river town of New Hope and the village of Lahaska, with its many antique dealers, tempting gift shops, and country restaurants. One of southeastern Pennsylvania's most popular attractions for oenophiles, Bucks Country reflects its founder's former show-business career as well. Arthur Gerold was the owner of Brooks-Van Horn Costume Company in New York City, the oldest and largest theatrical costumer in America, and in 1973 he established Bucks Country in a 150-year-old New Hope barn. Ten years after founding his winery, Gerold sold his costume business to devote full time to the expanding new enterprise.

A guided tour of Bucks Country Vineyards and Winery includes the basement-level wine cellars, the aging and bottling rooms, and a tasting. In addition to the tour, however, there are extras, one of which only some-one of Gerold's experience could provide. The commodious barn con-tains a newly installed Fashion and Wine Museum, a bakery and cheese shop, and a boutique. The Fashion and Wine Museum, located on the second floor, houses a rare collection of original costumes worn by famous Broadway stars. A collection of antique wineglasses (some of which were once in the collection of New York City's Metropolitan Museum of Art), exhibits on the champagne-making process and the history of wine, as well as many artifacts of Pennsylvania viticulture are also on display.

Bucks Country's bakery is located on the ground floor. Here croissants and French bread are freshly baked on weekends; homemade cheese can be sampled and purchased every day. The Grape Vine Boutique carries a large selection of wine-related accessories.

Tours of the winery are regularly scheduled on weekends. During the week, tours are self-guided, unless you make a prior appointment for a guide. There is a charge of $1 for the guided tour, which takes about an hour. (Half of this fee is donated to the Helen Hayes Scholarship Fund at the local Solebury School.) In summer, tours also take in the small demonstration vineyard which surrounds the winery. Afterwards, an in-formative tasting of nearly a dozen different wines is conducted in the ground-floor tasting/retail area.

Bucks Country Vineyards makes 75,000 gallons of wine annually from grapes furnished by contract growers in Pennsylvania. Ninety percent is still table wine, two percent *methode champenoise* sparkling wine, and eight percent fruit wine. A long product list includes more than twenty wines, among which are Blanc de Vidal, Eye of the Pheasant (a blush wine), Blanc de Blanc Champagne (a gold-medal winner), and Dutch Apple Wine.

Open daily all year: Monday-Friday, 11-5; Saturdays and Holidays, 10-6; Sunday, 12-6. Guided tours: regularly scheduled on weekends; self-guided, or by appointment, on weekdays. $1 charge for guided tours. Free tasting. Retail outlets: wine, wine-related accessories, homemade baked goods and cheeses. Wine and Fashion Museum. Limited picnic facilities (in the court-yard in front of the winery). Italian Wine and Music Festival (mid-July), Art Festival (mid-August), Harvest Festival (mid-September), Octoberfest (early October), and Nouveau Festival (weekend after Thanksgiving). Access for the handicapped (to the winery and tasting room). Directions: On right side of Route 202, 3 miles south of New Hope.

North East

Penn Shore Winery and Vineyards
10225 East Lake Road
North East, Pennsylvania 10428
Telephone: (814) 724-8688
Owners: George Luke, Blair McCard, and George Scieford

Two wineries in northeastern Pennsylvania near the shores of Lake Erie are instructive to visit as a pair since they complement each other as wine-touring experiences—Penn Shore Winery and Vineyards and Presque Isle Wine Cellars. Both are located in the heart of one of the finest grape-growing areas in the East, a region that also claims the largest planting of Concord grapes in the world.

The larger of the two wineries is Penn Shore (126,000 bottles annual-ly). One of the oldest and largest wineries in the state, it was founded in 1969 following the enactment of Pennsylvania's Limited Winery Act of 1968, which permitted wineries in the state to produce and sell their wines at the winery premises.

Penn Shore's vineyards surround the winery. Here 350 acres are planted to Seyval, Vidal, Ravat, Baco, Chancellor, Catawba, Delaware, and Niagara. The winery makes both premium award-winning white wines (its Signature Vintage label Ravat Blanc, Seyval, and Vidal fall into this category) and grapey, fruity, sweet northeast-style wines from its large planting of native American Labrusca grapes.

Guided tours of the winery are conducted in the summer. In winter, tours are self-guided, but a permanent exhibit of photographs of the winemaking process helps make up for the lack of a guide. Tours cover the whole production facility, following the path of the grape from harvesting (the winery's Signature wine grapes are picked by hand), to crush pad, fermenters, and the aging cellar. Tours end in the tasting room for a sample of the finished product.

In addition to its regular schedule of tours and tastings, the winery sponsors an annual wine festival, Wine Country USA, on the second weekend in September.

Open daily all year: Monday-Saturday, 10-6; Sunday, 12-5:30. Guided tours: June-September; self-guided: October-May. Free tasting. Retail outlet: wine, gifts. Wine Country USA Festival (second weekend in September). Limited access for the handicapped. Directions: From North East, north on Route 89 to Route 5. West on Route 5, 1½ miles to winery (on your left).

Presque Isle Wine Cellars
9440 Buffalo Road
North East, Pennsylvania 16428
Telephone: (814) 725-1314
Owners: Mr. and Mrs. Douglas P. Moorhead

Five miles west of Penn Shore Vineyards on Route 20 is tiny Presque Isle Wine Cellars. The winery is a good stop after a morning tour of Penn Shore, for not only does Presque Isle have picnicking facilities, but you will get a contrasting look at a small operation (10,000 gallons) and taste some special wines.

Presque Isle's size limits tours of its facility to small groups—a real plus. For if you do elect to take one of the informal tours, you will get to see all of the steps in wine production. But you won't get to see the winery's mainly vinifera and French hybrid vineyard unless you make a special appointment, for it is located a mile up the road at the Moorheads' farm. Clearly, the main reason most people stop here is to sample and purchase the excellent wines.

Presque Isle makes between fifteen and sixteen different wines a year, most of which can be characterized as dry. The line includes the varietals Cabernet Sauvignon, Cabernet Franc, Gewürztraminer, Chardonnay, Vidal Blanc, and Seyval Blanc, among others. Its Cabernet Sauvignon has been called one of the best in the East.

All the wines produced are available for tasting, and in addition to its wine, which is sold only on the premises, the winery sells cheese and winemaking supplies. (It has developed quite a reputation as a supplier of the latter.) If you have brought along a loaf of bread and perhaps some fruit to complement the wine and cheese, head down to Twelve Mile Creek, behind the winery, and enjoy a picnic on its banks. (There are more conventional picnic facilities next to the winery building.)

Open all year: Tuesday-Saturday, 8-5. Guided tours. Free tasting. Retail outlet: wine, cheese, winemaking supplies. Picnic facilities. Wine festival (latter part of September). Directions: On Route 20 between the towns of Harbor Creek and North East.

Stewartstown

Naylor Wine Cellars
Ebaugh Road
Stewartstown, Pennsylvania 17363
Telephone: (717) 993-2431
Owners: Richard and Audrey Naylor

Naylor is a seven-year-old winery in a new building (it outgrew its former potato cellar location) in southeastern Pennsylvania's York County a few miles north of Stewartstown. Owned by Richard and Audrey Naylor, it is located amid twenty-two acres of vineyards the couple established in 1975. Mr. Naylor is President of the Pennsylvania Wine Growers Association. Vineyard, planted to a number of grape varieties, produces 125,000 bottles annually. Eighty percent of the grapes needed for production are grown in the surrounding estate vineyards; the rest are purchased. Vineyards produce Vidal, Seyval, DeChaunac, Niagara, Catawba, Chambourcin, Chardonnay, Riesling, and Cabernet Sauvignon. The product line—ninety-five percent table wine, five percent fruit wine—consists of varietal wines made from the aforementioned grapes, as well as proprietary wines, such as York White Rosé, Rhinelander, and First Capitol. Complete tour, free tastings, many festivals, award-winning wines. Open all year: Monday-Saturday, 11-6; Sunday, 12-5. Guided tours. Free tasting. Retail outlet: wine, gifts. Festivals: weekends, June-October. Access for the handicapped. Directions: South on Interstate 83 to exit 1. East on Route 851, 4 miles to Stewartstown and Route 24. Take Route 24 north exactly 2 miles. Turn left to winery. (Mailing address: R.D. 3, Box 424, York, Pennsylvania 17403.)

# The South

Of the seven wine-producing Southern states discussed in this book—Alabama, Arkansas, Florida, Mississippi, North Carolina, South Carolina, and Virginia—the richest for wine-touring is Virginia, with twenty-five wineries of the approximately sixty in the area. But more important, Virginia is also the South's leading producer of premium wines made from vinifera grapes. With few exceptions, the rest of the South's vineyards are largely planted to its own special native grape varieties (belonging to the Muscadine species) and are sparsely sprinkled throughout the region.

Vitis Rotundifolia, the botanical name for Muscadines, was discovered by early explorers of the New World who wrote that its rich, fragrant scent, especially marked in September, was detectable for miles before they reached land. For four centuries Southerners have made wine from these grapes, earning for Muscadines the title of America's first truly original wine.

Unlike other native northern American cousins such as the Labruscas or European counterparts, Muscadines grow in clusters, not in bunches. Each berry is large and round, rather like a marble or cherry, and one vine has been known to produce as much as an acre of foliage from a trunk two feet thick. Fermented alone, Muscadines produced a rather harsh product, so Southern winemakers have traditionally sweetened these wines. Of the varieties grown, the best known and most widely planted is the native Scuppernong, with the more recently developed Muscadine hybrids Magnolia, Noble, Carlos, and Dixie supplanting Scuppernong in the newer vineyards.

Virginia was the first of the North American colonies to cultivate grapes, with attempts to grow vinifera dating back to 1619. George Washington planted a garden vineyard at Mount Vernon, but it was Thomas Jefferson who was the most ambitious of colonial winegrowers. He planted imported vinifera vines at Monticello and arranged for the services of a French winemaker. Jefferson's efforts

73

failed, however, largely because there were none of the modern sprays available to combat the pests and diseases that thrive in Virginia's moist, warm climate.

In the last ten years, thanks to modern science and farm winery legislation, Jefferson's vision of a Virginia wine industry is finally being realized. Vineyard acreage has increased dramatically in this period from less than 50 acres to more than 1,000. And family-run farm wineries are now being joined by large European-financed operations.

Many of Virginia's most rewarding and pleasurable wine touring experiences are to be found near the towns of Middleburg and Charlottesville. These are especially picturesque areas with plentiful restaurants, attractive overnight lodgings, and interesting historic sites. Surrounding Middleburg in the hunt country are Meredyth Vineyards, founded in 1972, and Piedmont Vineyards. Further south in the newly designated Monticello viticultural area is the Italian-owned Barboursville Vineyards with its all-vinifera plantings, the underground Montdomaine Cellars, and the latest winery to be bonded in the state, Oakencroft, on a registered Polled Hereford farm.

The remaining Southern states are more sparsely populated with wineries. Some of these, though, provide a refreshing experience for the traveler. Biltmore Estate's new winery is incorporated in the landmark property, its 250-room chateau being one of North Carolina's leading tourist attractions. Charming Truluck Vineyards Winery in South Carolina; Arkansas' big Wiederkehr "Swiss Village" consisting of winery, gift shops, restaurants, and extensive vineyard; and Florida's new Lafayette Vineyards dedicated to educating the public about Muscadines are some of the delights awaiting wine connoisseurs visiting the Old South.

ALABAMA

Perdido

Perdido Vineyards
Baldwin County Highway (Route 1, Box 20-A)
Perdido, Alabama 36562
Telephone: (205) 937-9463
Owners: Jim and Marianne Eddins

Perdido Vineyards holds Alabama's farm winery permit number 1. It was founded in 1979 by Jim Eddins, a practicing engineer, and his wife, Marianne, a trained biologist who works in the winery full time as winemaker. Perdido, the state's first bonded winery since the repeal of Prohibition, is its only one to date. Its location, thirty-five miles northeast of Mobile and only a couple of yards from Interstate 65 in rural Baldwin County, makes it a convenient stop for the traveler passing through.

The winery, constructed in the Spanish Mission style with white stucco facade and arches, was built specifically with tours in mind. Wings radiating from a central hub hold the different winery operations and make them readily accessible to the visitor. Perdido's vineyards surround the winery building, their 102 acres planted to the native grape varieties Magnolia, Scuppernong, and Higgins.

Tours of the winery include a short narrated slide presentation covering the yearly cycle of winemaking and viticulture; usually nine wines are available for sampling in the tasting room. Generally the choice includes four whites, three reds (one a rosé), and two apple wines. In addi-

tion, a wine-based beverage is offered. Depending on the season, it might be a hot mulled wine or a refreshing Sangria (recipes are freely distributed on request).

A complete tour of Perdido's winery, including a tasting of its product, should take around thirty minutes. And while there are no formal picnicking facilities, Jim Eddins invites you to make yourself comfortable on the winery's lawn or under one of its porticos overlooking the vineyards.

Perdido produces 13,000 cases of still table wine yearly. Its product line includes varietal Magnolia; various Muscadine generics, including the archly named and very popular Rose Cou Rouge (Red Neck Rose), the bottle of which comes outfitted with its own miniature straw hat, red bandanna, and Wallace button; and two popular apple wines (apples are a big agricultural crop in the state and forty percent of the wine consumed in Alabama yearly is apple-based).

Open all year, except July 4th, Thanksgiving, Christmas, and New Year's Day: Monday-Saturday, 10-5. Guided tours daily, include slide presentation. Free tasting. Retail outlet: wine, gifts. Picnicking permitted on grounds. Special events: pruning clinics (January); Cou Rouge Classic (November). Access for the handicapped. Directions: Interstate Highway 65 to exit 45 (Perdido-Rabun). Winery located just off exit on southeast side of highway on Baldwin County Road. (Approximately 45 minutes from Mobile, 2 hours from Montgomery.)

ARKANSAS

Altus

Wiederkehr Wine Cellars
Champagne Drive (Route 1, Box 14)
Altus, Arkansas 72821
Telephone: (501) 468-2611
Owners: The Wiederkehr Family

At the summit of St. Mary's, one of the Boston Mountains of Arkansas, is a cluster of buildings that looks as though they had been transplanted from the Swiss Alps. Called Wiederkehr Village, the group of European-style structures shelters the many facilities of Wiederkehr Wine Cellars.

The oldest and largest winery in the Southwest, Wiederkehr was founded in 1880 by Johann Andreas Wiederkehr and his wife, Katrina. The original cellar they dug for their wines, listed on the National Register of Historic Places, now houses the popular Weinkeller Restaurant. The cabin built above the cellar as the first Wiederkehr residence still stands. Housed in picturesque chalets throughout the village are the many winery facilities and a well-stocked gift shop.

Wiederkehr is now operated by the third generation of the Swiss family. Herman, son of Johann and Katrina, followed in their tradition of old-world winemaking. Today his sons, Leo and Alcuin, both trained in viticulture and oenology at UC Davis, have integrated modern technology and changing tastes while respecting their heritage, and their children are learning the ropes.

Guided tours of the modern winery take about thirty minutes; they cover the original cellar and log cabin, the modern cellars and production facility, and the tasting room, where a short film, supplemented by exhibits, covers the history of the winery. You're welcome to stroll along a path through the vineyards; it leads to an observation platform which offers beautiful views of the countryside below.

The Weinkeller Restaurant serves Swiss-German cuisine amid the atmospheric surroundings of the historic wine cellar; Die Trauben Stube (The Grape Lounge) is open for cocktails daily. A gift shop features merchandise from around the world; Wiederkehr wines are sold in a separate retail store.

At the edge of the complex are Wiederkehr's 350 acres of vineyards. Planted to native American grapes, French-American hybrids, and viniferas, they produce 2 million gallons of wine a year. Seventy-five percent of the harvest becomes still table wine, fifteen percent sparkling wine, and the rest fortified and fruit wines. Wiederkehr is particularly known for its Rosé de Cabernet Sauvignon, Muscato di Tanta Maria, and Hanns Wiederkehr Champagne.

Open all year: Monday-Saturday, 9-4:30. Guided tours of the winery; self-guided tour of the vineyard. Free tasting. Retail outlet: wine. Gift shop. Weinkeller Restaurant, Swiss-German cuisine, open for lunch 11-2, dinner 5-9, Monday-Saturday. Die Trauben Stube (cocktail lounge) open: 5-1 a.m., entertainment Wednesday-Saturday nights. Special events include Annual Festival (Weinfest) first two weekends in October, Christmas Tree Lighting Ceremony, New Year's Eve Celebration. Access for the handicapped. Directions: From Fort Smith, east on Interstate 40 to exit 41. Follow Road 186 and signs to Wiederkehr. (The winery is 40 miles east of Fort Smith.)

FLORIDA

Tallahassee

Lafayette Vineyards and Winery
6505 Mahan Drive (Route 7, Box 481)
Tallahassee, Florida 32308
Telephone: (904) 878-9041
Owners: C. Gary Cox and Gary M. Ketchum

Staunch, once-loyal consumers of California chardonnays have been seen leaving Lafayette Vineyards and Winery with cases of its semi-dry Muscadine wine in hand. But such disloyalty doesn't surprise Lafayette's owners, since one of the purposes of their Florida winery is to educate the public in the appreciation of Muscadine, a native southern grape which was thriving here long before Columbus set foot on the continent.

Founded in 1983, Lafayette's winery/tasting room is located east of the state capital on Highway 90. Fronting on the highway, its two-story building is faced with coquina (a masonry block material made from a mixture of limestone, seashell, and coral fragments). Spanish-style arches and shuttered windows frame Lafayette's thirty-eight acres of vineyards, planted largely to Magnolia and Welder, improved varieties of the native Muscadine. Stover, a French-American hybrid, is also planted, as well as numerous experimental grape varieties. The current annual output of the winery is 35,000 gallons, though the volume increases each year. Eighty-five percent of the product is still wine; the rest, sparkling wine.

The tour of Lafayette's working winery is self-guided weekdays; a guide is available on weekends. Either way the route is the same. A walkway suspended above the winery floor provides a view of the operating facility below. At the end of the walkway, a door leads to an outdoor balcony overlooking the vineyards.

The tour focuses on the history of the winery and offers a broader perspective on winemaking in the Tallahassee area. Tours conclude with a complimentary tasting in a comfortably appointed tasting room with a long stand-up oak bar. Picture windows on interior walls afford views of the winemaker's laboratory and the bottling line. Five wines are generally available for sampling daily, whether you take the tour or not. Antique display cases along the walls hold various wine-related gift items; these are interspersed with the Lafayette product, which is displayed in case lots. The winery has received early recognition for its semi-sweet white, a Muscadine called Lafayette.

While Lafayette Vineyards has no restaurant of its own, Cross Creek Restaurant, adjacent to the winery, features Southern-style cooking and might make a nice stop after your tour.

Open March-August; Monday-Saturday, 10-6; Sunday, 12-6; September-February: closed Mondays. Tours guided on weekends, self-guided weekdays. Free tasting. Retail outlet: wine, gifts. Access for the handicapped limited to tasting room. Directions: From downtown Tallahassee, winery is approximately a 15-minute drive east on Highway 90. Winery is located on your right about ¾ mile before Interstate 10.

MISSISSIPPI

Merigold

The Winery Rushing
Old Drew Road (P.O. Drawer F)
Merigold, Mississippi 39759
Telephone: (601) 748-2731
Owners: Diane and Sam Rushing

Established in 1977, The Winery Rushing is the first winery in Mississippi to be bonded since the repeal of Prohibition. Its 25-acre vineyard, one of the largest commercial plantings of muscadines in the state, is planted to the varieties Carlos, Magnolia, and Noble. The vineyard is across a bayou from the winery buildings, which rise out of the cotton fields on the sloping banks of the Sunflower River.

The Winery Rushing is owned and operated by Sam and Diane Rushing, who acquired their taste for wine while living in Europe. Twenty-minute tours of the small facility are structured to the interests of participants; tastings of the Rushing wines conclude your visit. From the native Southern muscadines in its vineyards, Rushing produces 6,000 cases of wine a year; all are estate bottled. The wines include varietals as well as four proprietary blends: Rushing Red, White, Rosé, and Sweet White. Both reds and whites have received national and international recognition, including a number of awards.

Located above the winery's cellar, Top of the Cellar Tea Room, operated by the Rushings, is furnished with tables, chairs, and pottery crafted by local artisans. The restaurant, open for lunch Tuesday through Saturday from 11:30 to 1:00, offers a choice of two entrées and an array of tempting

RUSHING
White

Mississippi Muscadine
Table Wine

Grown, Produced & Bottled by
The Winery Rushing
Merigold, Mississippi
Alcohol 12.0% by Volume
Bonded Winery Number One

desserts. The Rushings invite you to relax before lunch on the restaurant's deck, which overlooks the river, and to sip a complimentary glass of wine as you enjoy a selection of appetizers. After lunch you are encouraged to visit the adjacent gristmill; corn meal ground at the mill is from the family's farm (meal can be purchased at the winery).

Open February-December: Tuesday-Saturday, 10-5. Guided tours. Free tasting. Retail outlet: wine, gifts. Restaurant: Top of the Cellar Tea Room. Picnic facilities. Access for the handicapped. Annual Merigold Wine and Crawfish Festival (May). Directions: Highway 61 to Merigold; then east on Old Drew Road for 3 miles to winery (on your left).

Starkville

Thousand Oaks Vineyard & Winery
Old Highway 82 East (Route 4, Box 293)
Starkville, Mississippi 39759
Telephone: (601) 323-6657
Owner: Robert M. Burgin

Established in 1972. Rustic log tasting room surrounded by vineyard. Winery takes its name from the stately oak trees which grow nearby. Twenty-five acre vineyard: 12 acres planted to French-American hybrids, 12 acres to muscadines, 1 acre to viniferas. Ten different labels (eight varietals and two blends) include Villard Blanc, Baco Noir, Verdelet Blanc, Magnolia, Niagara, Noble, and Maréchal Foch. 25,000 gallons annually. Most noted for Magnolia (six awards in last four years). Open all year, except Thanksgiving, Christmas, and New Year's Day: Monday-Saturday, 10-4. Guided tours. Free tasting. (Any wines available for sale can be sampled.) Retail outlet: wine, gifts. Picnicking permitted on the grounds (no tables). Limited access for the handicapped. Directions: From Starkville take U.S. Highway 82 east to Hickory Grove Road exit. North at Hickory Grove Road to Old Highway 82; right on 82 to winery (one mile on your right).

NORTH CAROLINA

Asheville

The Biltmore Estate Wine Company
1 Biltmore Plaza
Asheville, North Carolina 28803
Telephone: (704) 274-1776
Owner: Mr. William A. V. Cecil

Biltmore House and Gardens, a 250-room French Renaissance chateau built by George Washington Vanderbilt on his 8,000-acre estate in the Blue Ridge Mountains near Asheville, is one of North Carolina's most popular tourist attractions. Now, in addition to the tour of the chateau, with its rich furnishings and elaborate gardens, you can also visit its recently opened winery.

The brainchild of William Cecil, president of the Biltmore Company and Mr. Vanderbilt's grandson, the winery has been in the works for ten years. It is an undertaking consistent with the splendor of the house and Mr. Cecil believes, with the vision of his grandfather's plan to develop the land. "At the time my grandfather was finalizing the landscaping plans for Biltmore, Europe was still in the throes of solving problems created by the devastating plant louse, phylloxera," Cecil says. "Knowing that

there was phylloxera at Biltmore, I'm sure there was reluctance to plant European wine grapes on the estate. But the introduction of vineyards and a winery is in keeping with my grandfather's intention to maintain a working estate in the European tradition."

The 30,000-square-foot production area of the winery is located in renovated dairy barns designed, as was the mansion, by noted architect Richard Morris Hunt. Centered around a clock tower, the complex incorporates a new visitors' center and tasting room. The center is lavishly decorated with imported Portuguese tiles depicting winemaking themes; stained glass windows originally commissioned from LaFarge by William Vanderbilt for his Fifth Avenue residence; and hand-applied wall stenciling, incorporating the Vanderbilt family crest. Tours of the winery are quite complete and well planned; as you would expect, the estate has been catering to throngs of tourists ever since it opened to the public in the 1930s. Upon arrival at the winery you are met in the welcome center by a guide who will give you a brief introduction to the facility, its history, and special features. You are then free to tour the winery at your own pace. Guides stationed along the route will explain the different activities taking place in the production areas and answer your questions. Balconies and interior windows provide interesting vantage points from which to view the tall fermentation tanks and champagne bottling room and to look into otherwise inaccessible areas such as the barrel cellars. Audio-visual presentations, a museum of wine artifacts, and the "Scholar's Walk" which contains displays on viticulture all enhance your visit. The tour ends in the winery's spacious tasting room where a structured tasting of one of Biltmore's wines is conducted (a broader sampling will probably be permitted as the winery's production increases).

Biltmore Estate Winery has 125 acres planted to vineyards; 11 to French-American hybrids, the rest to vinifera. The winery currently produces 35,000 cases of wine annually, including white, red, rosé, and *methode champenoise* sparkling wine. Grape varieties grown on the estate include Cabernet Sauvignon, Gewürztraminer, Chardonnay, Sauvignon Blanc, Merlot, Johannisberg Riesling, Gamay Beaujolais, and Pinot Noir. Biltmore is noted for its Sauvignon Blanc, Pinot Noir, Pinot Noir Rosé, and dry Riesling.

Tours of the winery alone cost $9.50; a ticket combining tours of both house and winery is $15, a tour that will take at least five hours. There are picnic facilities on the estate, under the pine trees next to its restaurant, Deerpark. Some of the tables are sheltered by an overhanging roof. If you prefer to dine more formally, Deerpark is open daily from 11 to 3.

Open all year: daily 10-5. Self-guided tours of winery, $9.50. Or combined with tours of Biltmore House, $15. (Children 11 or younger, accompanied by an adult, free). Museum. Film. Tasting (included in winery tour price). Retail outlet: wine and wine-related merchandise. Picnic facilities. Special event: Christmas at Biltmore. Accessible for the handicapped (elevators in all buildings and wheelchairs provided). Directions: Interstate 40 north to Route 25. Route 25 north 3 miles to estate (on your left).

Rose Hill

Duplin Wine Cellars
Highway 117 North (P.O. Box 756)
Rose Hill, North Carolina 28458
Telephone: (919) 289-3888
President: David Fussell

Duplin Wine Cellars is located in the heart of North Carolina's grape-growing region midway between Goldsboro and Wilmington. The winery, a farmer-owned cooperative, is located in a former warehouse building which was remodeled for its current use in 1975.

A large winery by eastern standards, Duplin has a 500,000-gallon capacity. Its production is devoted solely to wines made from the southern muscadines, which are grown on 600 acres in the area on land that the winery either owns or leases.

Duplin's output is almost equally divided between table and fortified wines; a small quantity of *methode champenoise* sparkling wine makes up the difference. The Duplin list includes the varietal wines Scuppernong, Carlos, and Magnolia, and seventeen generic and proprietary wines, including North Carolina Chablis, North Carolina Burgundy, North Carolina Soft White, and North Carolina Soft Red. Its Scuppernong and Magnolia are particularly noteworthy.

Tours of Duplin Wine Cellars begin with a short (ten minute) video on viticulture and winemaking. This presentation stresses the German-Swiss heritage of early settlers to the area whose style of winemaking is reflected in the wine produced at Duplin today. Afterwards you get a tour of the working winery, ending up in the tasting/retail area where you can sample the Duplin product. Usually six wines are available for tasting daily. There is also a small museum on the premises containing rare wine bottles and antique winemaking equipment used before the Civil War when the winemaking industry in the state was most prosperous.

Next to the winery is a simple picnic area. For more comfortable dining, Duplin recommends the Graham House Inn, twelve miles north of Rose Hill in Kenansville. A former stagecoach stop and winery built in 1840, the Inn has been completely restored. Its wine list, of course, features Duplin products.

Open all year: Monday-Saturday, 9-5. Guided tours. Free tasting. Retail outlet: wine, gifts. Museum. Grape Festivals (September). Picnic facilities. Access for the handicapped. Directions: On Highway 117, 40 miles north of Goldsboro (or 40 miles south of Wilmington) in Rose Hill.

SOUTH CAROLINA

Lake City

Truluck Vineyards
Route 3 (P.O. Drawer 1265)
Lake City, South Carolina 29560
Telephone: (803) 389-3400
Owners: Dr. James P. Truluck, Jr., and Jay Truluck

James Truluck's 100-plus acres of French-American hybrids and viniferas flourish between the fields of soy beans and tobacco which are traditional South Carolina low-country crops. In 1971, Dr. Truluck, a dentist, established his vineyards on only two acres and crushed his first vintages at home. Five years later, he built his lovely French Provincial-style tasting room and three-story winery with underground aging cellar. He has been steadily increasing the grape varieties grown, the list of wines produced, the capacity of the winery, and its facilities ever since. (A restaurant and inn are planned for the near future.)

Truluck is a small family-run operation. Dr. Truluck still practices dentistry four days a week and shares the management of the winery with his son, Jay, who is assisted with the pruning, harvesting, and crushing operations by siblings Bowen and Cac An. Jay started working in the vineyard at age eleven and has been involved ever since. He studied oenology at UC-Davis and Mississippi State and, since 1983, has been in charge of production.

The Trulucks' extensive vineyard is planted to Chambourcin, Ravat, Villard Blanc, Cayuga White, Verdelet, Golden Muscat, Cabernet Sauvignon, and approximately 150 other experimental varieties. The family has the largest planting of Chambourcin in the East, from which they have produced award-winning (in both national and international competitions) Red, Rosé, and Blanc. Truluck's other labels include Carlos, Ravat Blanc, Villard Blanc, Munson Red, Cayuga White, Seibel Blanc, Golden Muscat, Seyval Blanc, and Vidal Blanc. Current production is upwards of 15,000 gallons annually.

Tours of Truluck's stuccoed winery are short (about 15 minutes) and cover its crush pad, fermentation and storage room, bottling room, and wine cellar. Afterwards you can sample all of its wines in the tasting room;

then you might like to enjoy a picnic near a large lake in the vineyard.

Open all year, except Thanksgiving, Christmas, New Year's Day, and July 4th: Monday-Saturday, 10:30-5:30. Guided tours. Free tasting. Retail outlet: wine, gifts. Festivals: Renaissance (May), A Day in France (July), and Jazz (October). Picnic facilities. Limited access for the handicapped. Directions: From Lake City, north on Highway 378 to Godwin Welch Store. Right at store onto Route 3 East. Winery 2 miles on your right.

VIRGINIA

Barboursville

Barboursville Winery
Route 777 (P.O. Box 136)
Barboursville, Virginia 22923
Telephone: (703) 832-3824
Owners: The Zonin Family

Thomas Jefferson would be pleased with the winegrowing developments in Virginia: Barboursville Winery, only a stone's throw from his famous home, Monticello, is realizing his dream of growing vinifera grapes and producing premium prize-winning wines.

Barboursville Winery is located on an 850-acre estate, formerly known as the Barboursville Plantation, in the newly designated Monticello viticultural district. The property was the home of statesman James Barbour, who served as Virginia's governor from 1812 to 1814. It includes the picturesque ruins of Governor Barbour's mansion, designed by his neighbor Thomas Jefferson (it was gutted by fire on Christmas Day, 1884); a new winery which sits amid a cluster of older farm buildings; and an expanding thirty-five-acre all-vinifera vineyard planted in separate plots throughout the property.

The estate is owned by the Zonin family, whose Italian forebears have been prominent European wine producers since the 1820s. The Zonins were the first to grow viniferas commercially in Virginia. Their first vintage was in 1978; since then annual production has reached 81,000 bottles. The product list is devoted to vintage-dated estate-bottled table wines, primarily varietals made from the premium wine grapes Cabernet Sauvignon, Chardonnay, Gewürztraminer, Johannisberg Riesling, Merlot, and Pinot Noir. It also includes a generic wine, Rosé Barboursville.

Award Winners for Barboursville have been its Chardonnay, Johannisberg Riesling, Pinot Noir Blanc, and Cabernet Sauvignon.

Tours of the Barboursville Winery take about ninety minutes and include a look at the winery operation and its modern imported Italian equipment; its vineyards (a cannon is fired periodically to keep away the birds and deer); and the ruins of the Barbour mansion (the tour of this Virginia Historic Landmark is self-guided.) Afterwards there are free tastings of the Barboursville product. The whole line is available for sampling.

Open all year: Saturdays only, 10-2 and 2-5. Guided tours of winery and vineyard. Self-guided tour of ruins. Free tasting. Retail outlet: wine, gifts. Albermarle Harvest Wine Festival in October at the Boar's Head Inn, Charlottesville. Access for the handicapped limited to the winery and tasting room. Directions: From Charlottesville, Route 20 north for 16 miles to Route 678. Route 678 south for ½ mile to Route 777. Right on 777 to winery.

Carter's Bridge

Montdomaine Cellars
Route 720
Carter's Bridge, Virginia
Telephone: (804) 971-8947
President: Waldemar G. Dahl

From an architectural standpoint, Montdomaine Cellars, tunnelled into one of the rolling hills of Albemarle County, is probably Virginia's most unusual winery. Bonded in 1980, it is also one of the state's newest and one of its largest. Montdomaine's modern state-of-the-art facility is totally underground, a setting which maintains an ideal year-round temperature of 50 to 65 degrees Fahrenheit.

Nestled in the woods south of Charlottesville (in the newly designated Monticello viticultural district), the winery faces southeast toward one of its six all-vinifera vineyards. Totalling fifty-two acres, all are within five miles of the winery; they are planted to the grape varieties Merlot, Cabernet Sauvignon, Chardonnay, Johannisberg Riesling, Pinot Noir, and Sauvignon Blanc. From these grapes and additional purchased crops, Montdomaine's California-trained winemaker produces 20,000 gallons of table wine annually. Only vintage varietals are produced. Recent award-winning wines include the Merlot, Chardonnay, Cabernet Sauvignon, and Johannisberg Riesling.

Tours of Montdomaine Cellars take just under an hour and are tailored to the interests of the guests. A guide takes you through the vineyard and then inside the underground facility, where some of the most advanced winemaking equipment in the state is installed. Your visit concludes with a tasting of two of Montdomaine's wines. If you visit the winery during the harvest in September, you'll be invited to help. There are picnic tables under the trees in the woods surrounding the winery.

Open May-October: Wednesday-Sunday, 10-4. Other months by appointment only. Guided tours. Free tasting. Retail outlet: wine. Picnic facilities. Directions: From Interstate 64, take Route 20 south 13 miles to Route 720. West on 720 to winery. Mailing address: Route 6, Box 168A, Charlottesville, Virginia 22901.

Charlottesville

Oakencroft Vineyards & Winery
Route 5
Charlottesville, Virginia 22901
Telephone: (804) 295-8175
Owner: Mrs. John B. Rogan

Oakencroft Vineyards is Virginia's most recent winery to be bonded. Also in the Monticello viticultural area, it was founded in 1978 by Felicia Warburg Rogan, a transplanted New Yorker and wine collector. The winery is located only three miles from Charlottesville on her husband John's registered polled Hereford farm. (Mr. Rogan is the owner of the Boar's Head Inn in Charlottesville, a popular resort which is host to the Albemarle Harvest Wine Festival every fall.) A red barn near a small lake populated by wild ducks and Canada geese serves as the winery building. Completing the pastoral setting are the grazing Herefords, the estate vineyard, planted on a gently rolling hill, and views of the Blue Ridge Mountains in the background.

Oakencroft's fifteen-acre vineyard is planted to the grape varieties Chardonnay, Seyval Blanc, Cabernet Sauvignon, and Merlot. From these grapes Mrs. Rogan produces varietal estate-bottled vintage-dated table wines. Her first release, a 1983 Seyval Blanc, won a bronze medal at the 1983 Wineries Unlimited Competition in New York State. Since then a Chardonnay, Seyval Blanc, Cabernet Sauvignon, and Merlot have joined the product line.

Self-guided tours of Oakencroft are available on the weekends when the winery has regularly scheduled hours. At these times you are welcome to tour the winery and vineyard at your leisure and taste the wines.

Open all year: Saturday and Sunday, 10-4. Other days by appointment. Self-guided tours. Tasting $1 (refunded with purchase of wine). Retail outlet: wine, gifts. Directions: From Charlottesville take Route 29-250 Bypass to Route 654 (Barracks Road). West on 654, 3½ miles to winery (on your left).

Culpeper

Rapidan River Vineyards
Route 4 (P. O. Box 199)
Culpeper, Virginia 22701
Telephone: (703) 399-1855
Owners: Dr. and Mrs. Gerhard Guth

Rapidan River Vineyards, founded in 1978, is one of the largest wineries in the Monticello viticultural area. It was started by Dr. Gerhard Guth, who was influenced by the legacy of a colony of German settlers who in the early 1700s had planted European grapes along the Rapidan River. Although these early vineyards failed, they inspired Dr. Guth to import German wine experts, plant a vineyard, and build a winery. Located north of Charlottesville in Culpeper, the winery and winemaster's house are red brick Williamsburg-style buildings surrounding a formal courtyard furnished with picnic tables.

Fifty acres on the surrounding estate are currently under cultivation to grapes. This all-vinifera vineyard is planted to Gewürztraminer, Chardonnay, Johannisberg Riesling, and Pinot Noir. Rapidan's product list includes estate-bottled varietal wines of the aforementioned grape varieties in addition to *methode champenoise* sparkling wines made from a blend of Pinot Noir and Chardonnay. Rapidan is noted for its Rieslings, which have won numerous awards.

The winery is open for guided tours and tastings every day. The ceiling in its handsome tasting/retail area incorporates beams salvaged from a pre-Civil War hospital; an antique stone fireplace also adds charm to the room.

Open all year, except Christmas: daily 10-5. Guided tours. Free tasting. Retail outlet: wine and wine-related merchandise. Picnic facilities. Directions: From Fredericksburg, Route 3 west to Route 20. Southwest on 20 to Locust Grove and Route 611. North on 611, 7 miles to the vineyards.

Edinburg

Shenandoah Vineyards
Route 686 (Route 2, Box 323)
Edinburg, Virginia 22824
Telephone: (703) 984-8699
Owners: James and Emma Randel

Long recognized as one of the richest agricultural areas in the country, the lush, 200-mile-long Shenandoah Valley, which stretches between the Blue Ridge and Shenandoah Mountains in western Virginia, is rapidly becoming known as a premier grape-growing region. Shenandoah Vineyards was the first winery established on the valley's fertile floor. It is owned by New Jerseyans James and Emma Randel, who established its vineyard in 1977 and two years later remodeled the rustic two-story red barn to hold the winery. Shenandoah, located on Route 628 outside the town of Edinburg, sits amid forty acres planted to French hybrids and vinifera grapes. Varieties grown include Vidal Blanc, Seyval Blanc, Chambourcin, and Chancellor Noir. Additional grapes needed for production are purchased from local growers.

The winery produces 24,000 gallons of table wine annually. Its product line includes both generics: Shenandoah Rose, Shenandoah Rouge, and Shenandoah White; and varietals: Vidal Blanc, Chambourcin, Seyval Blanc, Cabernet Sauvignon, and Chardonnay. Award winners include Cabernet Sauvignon, Chambourcin, and Shenandoah Blanc.

Shenandoah is a particularly friendly place to visit. The Randels' dog will usually announce your arrival, and you are offered the choice of either a guided or self-guided tour (the guided tour costs $1 per person). Both are conducted like a treasure hunt: maps lead you through the vineyard and winery and finally, to the pot of gold—the wine in the tasting room.

All wines that are available for sale can be sampled in the lofty pine-panelled room. On weekends cheese and freshly baked bread are sold here along with the Shenandoah product. Picnic tables by the winery barn are shaded by a row of maples and a grape arbor.

Open all year: daily 10-6. Tours: 10-3, guided ($1) or self-guided (no charge). Free tasting. Retail outlet: wine, gifts. Harvest Festival (first Sunday after Labor Day). Picnic tables. Access for the handicapped. Directions: Interstate 81 to exit 71 (Edinburg). West on Route 675 to Route 686. North on Route 686 1½ miles to winery (on your left).

Hume

Oasis Vineyard
Route 1
Hume, Virginia 22639
Telephone: (703) 635-3103
Owner: Dirgham Salahi

The new French Provincial winery of Oasis Vineyards is located seven miles from Virginia's scenic Skyline Drive and an hour due west of the nation's capital amid the rolling hills and farmlands of Virginia's hunt country. One of the largest wineries in Virginia (40,000 square feet), Oasis boasts a 10,000-foot underground cellar and a tasting and banquet room which can accommodate over 500 people.

Oasis was established in 1977; in that year Jerusalem-born Dirgham Salahi, an amateur winemaker and wine enthusiast, and his wife, Corinne, planted some grape vines as a hobby. The couple began marketing their wines in 1980 and today have forty acres planted to Cabernet Sauvignon, Merlot, Pinot Noir, Sauvignon Blanc, Gewürztraminer, Chardonnay, Johannisberg Riesling, Seyval Blanc, Chelois, and Chancellor. From these grapes and additional purchased crops the winery produces 20,000 gallons of wine annually. The product line includes both varietals and a *methode champenoise* sparkling wine.

Tours of the winery include its modern facility and large underground cellar and two films (shown on request) on winemaking and wine appreciation. Tours end with a tasting of the Oasis product in the spacious second-floor tasting room with its large rustic stone fireplace. The entire Oasis line is available for sampling (except for the champagne, which is only opened for groups and for which there is a $2 charge per flute). There are ample picnic facilities. Indoors, tables cluster around the stone hearth in the tasting room, where generous windows provide views of the vineyard and the Blue Ridge Mountains beyond. Outdoors, they are placed near a pond. A snack bar/deli on the premises stocks cheeses and cured meats to complement the wines for visitors who don't bring picnic lunches.

Oasis is particularly known for its Sauvignon Blanc, Cabernet Sauvignon, Merlot, Chelois, and Champagne (a gold-medal winner).

Open all year, except Thanksgiving, Christmas, and New Year's Day: daily 10-4. Guided tour. Films. Free tasting. Retail outlet: wine, gifts. Picnic facilities (indoors and out). Snack bar. Annual festival (May). Access for the handicapped. Directions: From Washington, D.C., Interstate 66 west to Marshall. At Marshall, take second exit, Route 647 South. Follow Route 647 to Route 635. Go west on 635 for approximately 10 miles to Hume and the vineyard (on your left).

Middleburg

Meredyth Vineyards
Route 628 (P.O. Box 347)
Middleburg, Virginia 22117
Telephone: (703) 687-6277
Owners: Archie and Dody Smith

Meredyth Vineyards is little more than an hour's drive west of Washington, D.C., on the edge of the picturesque town of Middleburg, the unofficial capital of Virginia's hunt country. (Middleburg is a good central point for a weekend's outing. Its streets are lined with art galleries, saddleries, and antique stores, and there is comfortable lodging and good food to be had in its plentiful Colonial inns and taverns.) One of the state's most celebrated wineries, Meredyth is located on a former cattle farm in the foothills of the Bull Run Mountains, an area known as the Piedmont region.

Meredyth Vineyards was founded in 1972 by Archie Smith and his wife, Dody. (It was the first winery to be bonded under Virginia's Farm Winery Law of 1980, legislation which Archie Smith was instrumental in seeing made state law.) The high cost of raising cattle on their property had prompted the Smiths to look for a profitable agricultural crop in the early '70s. After a year and a half devoted to researching grape cultivation, Archie Smith began planting the pastures of the family homestead to vines. The first few acres he planted were experimental, but, as the grapes began to prosper, the Smiths began to increase vineyard acreage in proportion to their farmland.

The Smiths' daughter, Susan Meredyth Smith, has become the winery's marketing director; son Archie, a former teacher at Oxford University, is the winemaker. Their vineyard, planted primarily to French-American hybrids, has grown to fifty-five acres and sixteen grape varieties, including Seyval Blanc, Villard Blanc, de Chaunac, Maréchal Foch, Leon Millot, Aurora, Pinot Noir, Cabernet Sauvignon, and Chardonnay. From this crop, plus a small quantity of purchased grapes, Meredyth produces 10,000 cases of wine annually. The product list includes the red wines de Chaunac, Cabernet Sauvignon, Maréchal Foch, and Villard Noir; whites Seyval Blanc, Chardonnay, Villard Blanc, Riesling, and Aurora Blanc; and Rougeon Rosé. All wines are estate bottled and vintage dated. Meredyth is particularly known for its Seyval Blanc, Chardonnay, Johannisberg Riesling, and Cabernet Sauvignon.

Tours of the winery, a former three-stall horse stable with a new mirror-image addition, take about forty-five minutes and thoroughly cover the working facility. It concludes with a tasting of between five and six of Meredyth's wines. The winery has an especially nice picnic area located at the highest spot on the farm, the site of some unidentified ruins. The vineyards surround the immediate picnic site, and the views beyond are lovely.

Open all year, except New Year's Day, Easter, Thanksgiving, and Christmas: daily 10-6. Guided tours. Free tasting. Retail outlet: wine, gifts. Picnic area in the vineyard. Special tastings monthly (contact winery for times and dates). Directions: From blinker in Middleburg town center, south on Madison Street 2½ miles to Route 628. At Route 628, go right 2½ miles to winery entrance (also on your right).

Piedmont Vineyards & Winery
Route 626 (P.O. Box 286)
Middleburg, Virginia 22117
Telephone: (703) 687-5528
Owner: Mrs. Thomas Furness

Piedmont Vineyards and Winery, situated on historic Waverly Farm, whose gracious antebellum mansion dates to the mid 1700s, is three miles south of Middleburg and only minutes from Meredyth Vineyards. The mansion, a registered Virginia Historic Landmark, is the home of octogenarian Mrs. Thomas Furness, who is often referred to as the matriarch of the winery. Mrs. Furness founded the business in 1973 as a way to preserve the land. Although neighbors, including the wine experts at nearby Virginia Polytechnic Institute, doubted the feasibility of growing

viniferas (citing Thomas Jefferson's failure 200 years before), the indomitable Mrs. Furness persevered—she went to school, read voraciously, hired an expert—and succeeded. "In Thomas Jefferson's time they didn't spray," she remarks. "I also thought the climate was right, not too hot and not too cold."

Thirty acres are now planted to Chardonnay, Seyval Blanc, and Semillon. And the winery's annual output has reached 10,000 gallons. The daily winery operation and vineyard management are overseen by Mrs. Furness's daughter, Mrs. William Worrall; her grandson, William E. S. Worrall; and UC Davis-trained winemaker Curtis Sherrer. Piedmont Vineyards makes four varietal white table wines, all of which are vintage dated.

The winery is located in a converted dairy barn on the property. Tours of the premises take about thirty minutes and are frequently guided by Mrs. Worrall or Mr. Sherrer, who cover the history of the winery, explain the vinification techniques, viticultural practices, and the working facility. Tours conclude with a tasting of the entire Piedmont line in the winery's combined retail/tasting area where you can purchase, among its other vintages, its semi-sweet Seyval Blanc, which is only available on the premises. Piedmont is particularly known for its Chardonnay, which has won several awards.

Open all year: Tuesday-Sunday and major Monday holidays: 10-4. Tours, both guided and self-guided. Free tasting. Retail outlet: wine, gifts. Picnic facilities. Access for the handicapped. Directions: From Middleburg, south on Route 626, 3 miles to winery (on your right).

Oak Grove

Ingleside Plantation Vineyards
Route 3, (P.O. Box 1038)
Oak Grove, Virginia 22443

Telephone: (804) 224-7111
Owners: The Flemer Family

Thirty miles east of Fredericksburg is Virginia's largest farm winery, Ingleside Plantation Vineyards. Ingleside shares a 2,000-acre estate with one of the largest landscape nurseries in the East. The land, currently overseen by Mr. and Mrs. Carl Flemer, Jr., and their sons, Douglas and Fletcher, has been in their family since 1890. Ingleside's extensive holdings include 1,000 acres in nursery production, over 100 greenhouses, several historic buildings, and a 35-acre vineyard.

The Flemers' interest in grape propagation came as a natural extension of their nursery business, but their European winemaker came to them purely by chance. Upon retiring as professor of oenology at Brussels University, Belgium-born and French-trained Jacques Recht sailed his catamaran across the Atlantic. Reading James Michener's *Chesapeake* on the voyage inspired him to sail into the bay and up the Potomac. On this fortuitous trip he met the Flemers, who were just building their winery and were looking for a winemaker. Recht has been at Ingleside ever since producing award-winning wines.

Ingleside's vineyard is planted to nearly thirty grape varieties, but primarily to Cabernet Sauvignon, Chardonnay, Johannisberg Riesling, Seyval Blanc, Vidal Blanc, and Chancellor. From these grapes, the winery produces 12,000 cases of wine annually—ninety-five percent still table wine, five percent sparkling wine. The product list includes the varietals Cabernet Sauvignon, Chardonnay, Johannisberg Riesling, Seyval Blanc, and Chancellor; and generics Roxbury Red, Wirtland Rosé, Ingleside White, (named for the landmark houses on the estate), and Fraulein. The winery has received many awards for its wines; its Chardonnay and Cabernet Sauvignon are particularly noteworthy.

Tours of the Ingleside winery, located in a converted dairy barn, take about half an hour and cover the production of its still table wines. (By appointment large groups are shown the temperature-controlled champagne production area.) Afterwards you can sample six or seven of Ingleside's wines in the barn's loft/tasting room. (Generally the whole line can be sampled except for the champagne and special vintages.) Afterwards you can stroll through the gardens and grounds of the handsome estate and picnic near the vineyards, where tables are provided.

Open all year: Monday-Saturday, 10-5; Sunday, 1-5. Guided tours. Free tasting. Retail outlet: wine, gifts. Picnic facilities. May Festival (third weekend in May): arts, crafts, music, food, and wine. Limited access for the handicapped. Directions: From Fredericksburg, east on Route 3, 30 miles to winery.

The Midwest

The Midwestern states of Indiana, Michigan, Minnesota, Missouri, Ohio, and Wisconsin are a diverse lot in relation to grapegrowing history and wine production. They range from Missouri and Ohio, where the industry is making substantial progress in restoring its important pre-Prohibition status, to Minnesota and Wisconsin, where wine grapes grow against all odds and wineries are few.

Wisconsin, largely a fruit state, has some excellent fruit wineries which produce wines from local cherries, cranberries, and apples, but only one estate grape winery—century-old Wollersheim Winery near Prairie du Sac on the Wisconsin River, and founded by Agoston Harazathy who later made California wine history.

Minnesota's bitter winters also preclude extensive grapegrowing. Here the vines are frequently cut free from their trellises and mulched on the ground during winter in order to survive the sub-zero temperatures. The rewards of such dedication can be enjoyed at Alexis Bailly's log cabin winery and vineyard outside Hastings and at the Northern Vineyards winery in Stillwater, which makes wine from grapes provided by the Minnesota Winegrowers Cooperative. In both Wisconsin and Minnesota, French-American hybrids are predominantly grown as are the winter-hardy hybrids Kay Gray, St. Croix, and Swenson Red, developed by Wisconsin grape breeder, Elmer Swenson.

Indiana, Michigan, Missouri, and Ohio are the most important Midwestern winegrowing states, and consequently the most interesting for tourists. Indiana, the smallest of the four, has eight wineries, all found in the south. At Bloomington is Oliver Wine Co. and nearby Possum Trot Vineyards. But the state's largest and most favorable viticultural region is the Ohio River Valley area. Here you will find the big Hoosier Farm winery, Huber Orchard in Borden, and excellent St. Wendel's chalet-style facility near Wadesville. All of Indiana's wineries were founded after the passage of a farm winery bill in 1971. The vineyards are predominantly planted to French-American hybrids.

Michigan has eighteen wineries, 13,000 acres planted (only a fraction of which are used for wine), and two viticultural areas, Leelanau and Fennville. Grapes grow easily along the Lake Michigan shore. The southwestern region, around Paw Paw, is largely planted to Concords, which account for roughly eighty-eight percent of Michigan's production, most of which is used for jelly, juice, and jam. Conveniently located next to each other in Paw Paw are the state's two biggest wineries, Warner Vineyards and St. Julian.

In the northwest, the Lake Leelanau and Grand Traverse Bay areas are largely planted to wine grapes. Here you will find the premium Tabor Hill Vineyard in Buchanan, with its gourmet restaurant and cross-country ski trails which wind around the vineyard blocks. Good Harbor Vineyards, Chateau Grand Traverse, and Boskydel are located in the Leelanau peninsula vacation area.

Missouri has twenty-four wineries and 2,000 acres planted to vineyards—predominantly Catawba and Concord—and two viticultural areas, Augusta and Hermann. Many wineries were founded here in the mid-1800s by German and Italian immigrants, including Stone Hill in Hermann, which at one time was the largest single American wine producer. Hermann is also the home of Hermannhof Wineries. Both facilities are national historic sites. Further southwest you will find the traditional Rosati Winery, founded by Italians, and the largest winery in the area, St. James.

Ohio, the most important Midwestern wine state, has forty-five wineries, approximately 2,600 acres planted to grapes, and three viticultural areas—Lake Erie/Chautauqua, Ohio River Valley, and the Lake Erie Wine Islands. Despite a curious state law which requires wineries to charge for tastings, Ohio offers some of the most agreeable touring experiences in the Midwest. Among them are the fascinating Lake Erie wine islands of Middle and South Bass, the sites of several historic wineries. On the mainland is the Mon Ami Wine Co., and twenty minutes east, Mantey, one of the largest Ohio producers.

INDIANA

Bloomington

Oliver Wine Company
8024 North Highway 37
Bloomington, Indiana 47401
Telephone: (812) 876-5800
Owners: The Oliver Family

Oliver Wine Company is housed in a cluster of buildings on Highway 37 eight miles north of Bloomington. Founded by William and Mary Oliver in 1972, it was the first small winery to open in the state. (Bill Oliver, professor of law at Indiana University in Bloomington, was instrumental in the drafting and passage of the state's small winery act in 1971.)

The winery, popular with nearby university students, produces 25,000 gallons of wine annually. Its site overlooks the valley where its fifteen-acre vineyard is planted to Cascade, Baco, and Vidal. Additional grapes needed for production are purchased. The winery's product line includes varietal, generic, and jug wines. Especially popular are its Camelot Mead (a honey-based wine sold in eight states and exported to Hong Kong and Singapore) and award-winning Soft Red, a proprietary blend.

Guided tours of the winery (available on the hour each Saturday and during the week by appointment) cover its barn-like production facility and adjoining underground cellar, which is built into the side of a hill. Afterwards you can taste all twelve of the Oliver wines in the nearby fieldstone building which houses the tasting room/retail area. Its interior is attractively finished with rough-hewn siding, a fieldstone fireplace, and oak furnishings. If you wish, you can meander down the hill to the creek or stop along the way for lunch at one of the convenient picnic tables. If you're lucky, you might spot the locally famous Oliver hot-air balloon, piloted by Bill Oliver, Jr.

Open all year: Monday-Saturday, 11-6; Sunday, 12-5. Guided tours regularly scheduled on Saturdays. Other days by appointment. Free tasting. Retail outlet: wine, gifts, cheese. Camelot Wine Festival (first weekend in June). Picnic tables. Directions: From Bloomington, 8 miles north on Highway 37.

Borden

Huber Orchard, Winery & U-Pick Farm
Route 1 (P.O. Box 202)
Borden, Indiana 47106
Telephone: (812) 923-9463
Owner: Huber Orchards, Inc.

Huber Orchard, Winery & U-Pick Farm is one of those rare wine places that can honestly be recommended to families with small children. Located in the rolling rural countryside of southern Indiana's Ohio River Valley, the winery is part of a large (450-acre) picture-perfect farm that includes, in addition to its grape-growing and winemaking activities, extensive crops that ripen throughout the season and that you can pick yourself.

So while the adults take the tour of the working winery, housed in a restored vintage-1938 dairy barn, older children can be more happily engaged in a game of tag on the spacious lawns surrounding the white-washed farm buildings, hop aboard a tractor-pulled hay wagon for a ride through the orchards, help pick a seasonal crop in one of the fields, or feed lunch leftovers to the friendly geese and ducks paddling in the lake. And if they're too little to be left alone, you can accompany them on these outings after the tour.

Huber Orchard Winery was founded in 1978 by brothers Gerald and Carl Huber and their wives, Mary Jeanne and Linda, on the ancestral farm that has been in the Huber family for six generations. The winery produces seventeen different wines, twelve grape and five fruit and berry, all of which are fermented from fruits harvested on the farm. Huber makes varietal grape wines from its French-American and Labrusca vineyard, which is planted to Chancellor, Chelois, Catawba, Concord, Seyval Blanc, Vidal Blanc, Aurora, and Niagara. These grape wines account for seventy percent of its 18,000-gallon annual output. Fruit wines derived from strawberries, blackberries, raspberries, peaches, and apples make up the remaining thirty percent.

Tours of Huber Orchard Winery are self-guided every day except Saturday, when there is a guide to show you the facilities, including the underground wine cellar. There is a handsome tasting room and retail area with a deli-style restaurant where you can purchase sandwiches, cheese, sausage, crackers, bread, and even pizza. The repast can be enjoyed at one of the numerous tables and chairs provided both indoors and out. After you have completed the wine tour and tasting, you can pick your own strawberries, peaches, apples, and numerous vegetables in season, cut your own Christmas tree, or select a perfect Halloween pumpkin from the pumpkin patch.

Since 1979, Huber Orchard Winery has received over fifty awards for its winemaking efforts. It is best known for its many fruit wines.

Open all year. Monday-Saturday: November-April, 9-5; May-October, 9-6. Sundays: January-April, 12-5; May-October, 10-6; November-December, 10-5. Guided tours, Saturday only; self-guided other days. Tasting: 20¢ per one-ounce sample. Retail outlet: wine, gifts, fruit, vegetables. Wine Festival (Labor Day weekend). Deli. Picnic facilities: indoors and out. Limited access for the handicapped. Directions: From Louisville, Kentucky, take Interstate 64 west and north to Greenville exit (119). At exit follow Highway 150 4 miles to Navilleton Road (across from Floyd County Bank). Turn right at this intersection and follow orchard and winery signs for 6 miles.

St. Wendel

St. Wendel Cellars
10501 Winery Road
St. Wendel (Wadesville), Indiana 47638
Telephone: (812) 963-6441
President: William Koester

In the rolling farmland of southwestern Indiana (west of Huber Orchard Winery, but also in the Ohio River Valley viticultural area) is St. Wendel Cellars, formerly called Golden Rain Tree Winery. Founded in 1975 by a group of local investors, the winery is one of the most modern in the state and produces many of its best wines.

St. Wendel has an underground cellar carved into the side of a hill, and a Swiss-Chalet-style tasting-room/banquet hall beyond. The main floor of the chalet is utilized as a combined wine tasting/retail area, which stocks home winemaking supplies and local crafts, as well as the St. Wendel product. A small deli located here carries lunch fixings such as cold cuts and cheese; the room's upper story is reserved for banquet facilities.

Lawns surround the winery; and you are welcome to spread a blanket and picnic. A small vineyard is planted on the property, but most of St. Wendel's vineyards are scattered within a ten-mile radius of the winery. These local vineyards are planted to the French-American hybrids Seyval Blanc, Vidal Blanc, de Chaunac, and Chelois, from which St. Wendel's Hindu winemaker, Murli Dharmadkari, produces upwards of 20,000 gallons of wine annually. (Dr. Dharmadkari holds a degree in viticulture and oenology from Ohio State University.) St. Wendel's product line includes both varietal and proprietary wines. Of its seven labels, the bestsellers are Spirit of '76 and Criterion White.

Short (fifteen- to twenty-minute) guided tours of the winery are available every day except Monday. These cover the underground cellar and winemaking facilities and conclude with a free tasting. (You are permitted to taste the whole line, if you like.)

Open all year: Tuesday-Sunday, 11-6. Guided tours at 11, 2, and 4. Free tasting. Retail outlet: wine, home winemaking supplies, crafts. Deli: cheese, meats, etc. Wine Festival (first weekend in June). Picnicking on the lawn (bring your own blanket). Directions: From Wadesville, take Wadesville-St. Wendel's Road to Luig's Road. South on Luig's Road (also known as the Blairsville-St. Wendel's Road) to Winery Road and St. Wendel's Winery (on your left).

Unionville

Possum Trot Vineyards
8310 North Possum Trot Road
Unionville, Indiana 47468
Telephone: (812) 988-2694
Owners: Ben and Lee Sparks

"Mom and pop" operation in hills of beautiful Brown County. Established in 1978. Attractive rural setting near Lake Lemon described by owner Ben Sparks as "a sylvan glen." Winery in a century-old barn. Tasting and retail area in adjacent bay-windowed fieldstone structure. Short, but very complete, demonstration of winemaking tools and equipment with commentary by the owners. All grapes purchased. Production: upwards of 8,000 bottles of vintage varietals and generics annually: Maréchal Foch, Vignoles, Seyval

Blanc, Vidal Blanc, Aurora, and Festival White, Zaragueya Sangria, and mulled wine. Best known for varietal wines (reds aged in American oak). Open daily: March-December, 10-7. Other months by appointment. Guided tours. Tasting: $1. Retail outlet: wine, gifts. Festival (first weekend in October.) Picnicking. Access for the handicapped. Directions: From Bloomington, State Road 45 to Trevlac. From Trevlac west on North Shore Drive 2.1 miles to Possum Trot Road. Right on Possum Trot to winery (on your right). (A visit to Possum Trot is good to combine with a stop at its Bloomington neighbor, Oliver Wine Company.)

MICHIGAN

Buchanan

Tabor Hill Winery
Mt. Tabor Road (Route 2, Box 720)
Buchanan, Michigan 49107
Telephone: (616) 422-1162
Owner: David F. Upton

Since their founding in 1968 by local businessman David Upton, Tabor Hill's vineyards have produced award-winning wines, some of which have been served in the White House. Tabor Hill's winery is located in the southwestern part of Michigan, just five miles from the shores of Lake Michigan. In addition, the winery operates tasting rooms in other areas of the state; these might be convenient if your time is limited.

A visit to the winery, however, is most rewarding. Here you can see how premium wines are made, sample the Tabor Hill product, walk through the vineyards, dine at a lovely restaurant, and even, in season, cross-country ski on the trails that circle Tabor Hill's surrounding vineyards.

Tabor Hill was among the first wineries in Michigan to cultivate vinifera grapes. In its twenty-acre estate vineyard are grown Chardonnay, Riesling, Pinot Noir, Vidal Blanc, Seyval, and Baco Noir. This vineyard produces twenty-five percent of the winery's needs; remaining grapes are purchased locally. Tabor Hill produces 90,000 gallons annually. Still table wines make up the bulk of this production, sparkling wines the remainder. Tabor Hill has fifteen labels, among them proprietary blends cuvée rouge, cuvée blanc, and cuvée rosé (a blue-ribbon winner); varietals Baco Noir Reserve,

Cabernet Sauvignon, American Chardonnay, and Johannisberg Riesling; and Vidal Demi Sec Champagne. The winery is particularly known for its Vidal Blanc Demi Sec, Chardonnay, and Riesling.

Tours of the modern winery, constructed of wood salvaged from an 1861 barn, include a look at its vineyard, cellar, and production areas, and conclude in the tasting room. Tours take about thirty minutes; after the tasting you can adjourn to the adjacent restaurant for lunch or dinner. In summer months there is dining on a screened porch overlooking the vineyard; in winter, tables are set near a glowing hearth.

Open all year, except Christmas and New Year's. April-December: Monday-Saturday, 11-5; Sunday, 12-5. January-March: Friday and Saturday, 11-5; Sunday, 12-5. Guided tours: April-December, 11-5. Free tastings. Retail outlet: wine. Restaurant (reservations suggested), open all year. Call for brunch, lunch, and dinner hours. Special events: May Festival, Jazz Festival (July), Chicken BBQ (August), and Harvest Festival (September). Cross-country skiing and instruction. Access for the handicapped. Directions: Interstate 94 to exit 16 (Bridgman). Follow signs 6 miles to the winery.

Fennville

Fenn Valley Vineyards
6130 122nd Avenue
Fennville, Michigan 49408
Telephone: (616) 561-2396
Owner: William Welsch

If you travel north from Bridgman on Interstate 94 and branch off on 196, which follows the shoreline of Lake Michigan, you'll come to Fenn Valley Vineyards. Like its neighbor to the south, Tabor Hill, it is in the southwestern part of the state, an area whose climate is moderated by its proximity to the lake. Family-owned-and-operated, the winery is located on a 230-acre farm in Michigan's first federally approved viticultural area, Fennville. It was established in 1973 following an extensive search by its owner, William Welsch, for a suitable vineyard site in the eastern United States. (This area of Michigan, with its rolling hills, light sandy soil, and moderate climate, has been favorably compared with the classic winegrowing regions of northern Europe.)

Fenn Valley has fifty acres planted to the viniferas Johannisberg Riesling, Gewürztraminer, and Chardonnay, as well as to winter-hardy French-American hybrids: Seyval Blanc, Vidal Blanc, Vignoles, Chancellor, and

Maréchal Foch. From these grapes and from its seven-acre peach orchard, Fenn Valley produces 20,000 gallons of wine annually. Ninety-five percent of this output is still table wine; the balance, fruit and sparkling wine. All wines are vintage-dated and estate-grown and bottled. Fenn Valley's list includes both varietal and generic wines. Among the former are Vidal Reserve, Dry Vidal, Seyval, Vignoles, Dry Foch, Foch Rosé, and Dry Foch Rosé. Generics are Vin Blanc and Ruby Red. Chardonnay and top-of-the-line Seyvals are aged in French and American oak. The winery is best known for its Vidal Blanc, Seyval Blanc, and Late Harvest Vignoles.

Tours of Fenn Valley's modern winery are self-guided. Signs posted on an observation deck over the wine cellar explain the equipment and cellar operations below; a short slide presentation on winemaking is shown continuously throughout the day. There are picnic tables on the grounds. Cheese and sausage are available for sale in the retail area, in addition to the wine (which you can have personalized with your own labels), gourmet wine vinegars, and (in season) fresh grapes and juice.

Open all year: Monday-Saturday, 9-5; Sunday, 1-5. Self-guided tours (including slide presentation). Free tasting. Retail outlet: wine, home winemaking supplies, gifts, food specialties. Picnic facilities. Directions: I96 to exit 34 (Fennville). East on M-89 for 3½ miles to 62nd Street. South on 62nd Street one mile to 122nd Street. East on 122nd Street ¼ mile to winery (on your right).

Fenton

The Seven Lakes Vineyard
1111 Tinsman Road
Fenton, Michigan 48430
Telephone: (313) 629-5686
Owners: Harry and Christian Guest

Michigan's newest winery. Owned and operated by father and son Harry and Christian Guest. (Son, Chris, is the winemaker.) Located in southeastern part of state, 2 miles from Fenton, near Flint. Winery takes its name from Seven Lakes State Park across the way. Twenty acres planted primarily to French-American hybrids. 5,000 gallons annually. Mostly still table wine, smaller amounts of fruit wine. Noted for Vignoles and de Chaunac Rosé. Especially comprehensive tour (1½ hours) covers vineyard and production facility. Slide presentation on winemaking. Open all year, except major holidays: daily, 10-5. Guided tours. Free tasting. Retail outlet: wine, gifts. Picnic facilities. Directions: Interstate 75 to Grange Hall Road. Grange Hall

Road west to Fish Lake Road. Fish Lake north to Tinsman Road. West on Tinsman to Seven Lakes Vineyard.

Lake Leelanau

Boskydel Vineyard
County Road 641 (Route 1, Box 522)
Lake Leelanau, Michigan 49653
Telephone: (616) 256-7272
Owner: Bernard Rink

Small (7,000-gallon) family-owned-and-operated winery, in northwestern corner of the state, in the federally recognized Leelanau region. Twenty-acre vineyard planted to Vignoles, Seyval Blanc, de Chaunac, Cascade Noir, Aurora Blanc, and Maréchal Foch. All wines vintage-dated and estate-bottled. Seven labels. Best known for Vignoles, de Chaunac, and Seyval Blanc. Modern winery, surrounded by pine trees, in especially scenic setting on a hill overlooking Lake Leelanau. Open all year: daily, 1-6. Short guided tours (10 minutes) regularly scheduled in summer between 1:30 and 5:30. Tours other months by appointment or by chance. Free tasting. Retail outlet: wine. Access for the handicapped. Directions: From Traverse City north on 633 to 641. North on 641 to winery. (Winery 15 miles from Traverse City.)

Good Harbor Vineyards
M-22 (Route 1, Box 891)
Lake Leelanau, Michigan 49653
Telephone: (616) 256-7165
Owners: John W. and D. Bruce Simpson

Another small (9,000-gallon) family-owned-and-operated winery, just south of Boskydel Vineyard. Located ½ mile from Good Harbor Bay on Leelanau Peninsula near Sleeping Bear National Park. The Simpson family, primarily cherry growers, has been involved in agriculture in America since the 1700s. Winery established in 1980. Twenty acres planted to Vignoles, Seyval Blanc, Riesling, Chardonnay, and Aurora. Varietal wines: vintage-dated and estate-bottled, Lake Leelanau appellation of origin. Generics: vintage-dated with Michigan appellation. Two-thirds of production still table wine; one-third fruit wine. Noted for Vignoles, dry Seyval Blanc, and cherry wine. Winery located behind the family farm stand/bakery. Open all year. May-October:

Monday-Saturday, 11-6; Sunday, 12-6. November-April: 11-6. Self-guided tours. Free tasting. Retail outlet: wine, gifts. Access for the handicapped. Directions: 3 miles south of Leland on M-22.

Paw Paw

St. Julian Wine Company
716 South Kalamazoo Street (P. O. Box 127)
Paw Paw, Michigan 49079
Telephone: (616) 657-5568
President: David Braganini

St. Julian, Michigan's oldest winery, was founded in 1921 by Mariano Meconi in Windsor, Ontario, as the Italian Wine Company. After Prohibition ended, Meconi moved his operations across the river to Detroit. And in 1936, he relocated his winery to Paw Paw, the heart of Michigan's grape-growing district. The center of Meconi's wine production, owned and operated by third-generation family members, has remained here ever since, though winery branches and tasting rooms have been added at other Michigan locations.

St. Julian produces 100,000 cases of wine annually from 12,500 acres of grapes supplied by local growers. Its product line includes still table wine, both *methode champenoise* and Charmat (bulk) process sparkling wine, fortified wine, fruit wine, and sparkling juices. Among its labels are Seyval Blanc, Vidal Blanc, Vignoles, Chancellor, and Niagara, Brut Champagne, Spumante, White Label Champagne, and Solera Cream Sherry. St. Julian is particularly noted for its Niagara, Friar's Blanc, Solera Cream Sherry, Chancellor Noir, and its reserve wines.

Tours of St. Julian are very informative and hospitable. The winery was the first in the state to offer free tastings and tours, which an average of 250,000 people now participate in each year. But its new winery (the original burned to the ground in 1972) and spacious tasting room are well equipped to handle the traffic. Guided tours begin with a glass of wine, and include a look at the grape crushing area, wine cellar, champagne area, bottling area, wine-processing area, and vat storage. At each stop, a full explanation of the production process is given and ample time allowed for questions. Afterwards there is tasting of the St. Julian product in the hospitality room. The combined tour and tasting takes about one and a half hours. Two picnic tables are located in front of the winery near a large antique press.

Open all year except Thanksgiving, Christmas, New Year's, and Easter:
Monday-Saturday, 9-5; Sunday, 12-5. Guided tours throughout the day (last
tour at 4). Free tasting. Retail outlet: wine, gifts, cheese and crackers. Wine
and Harvest Festival (September). Limited picnic facilities. Access for the
handicapped. Directions: Interstate 94 to Paw Paw (exit 60). At exit, go two
blocks north to winery.

Warner Vineyards
706 South Kalamazoo Street
Paw Paw, Michigan 49079
Telephone: (616) 657-3165
President: James J. Warner

No trip to Michigan wineries would be complete without a stop at Warner
Vineyards in Paw Paw. Located on the banks of the Paw Paw River within
walking distance of St. Julian Wine Company, Warner is the largest winery
in the Midwest and has some of the area's most attractive visitor facilities.
A small footbridge leads to Warner's visitor center, which consists of an
historic brick tasting room, formerly the town waterworks, and a 1912
Grand Trunk Railroad car parked alongside. Forty-five-minute tours of
the winery begin on the hour. First you'll see a fifteen-minute slide presentation in the tasting room, then proceed to the working winery located
directly behind it. After the tour and tasting you can relax in Warner's
Wine Garden Restaurant, located on the banks of the river. Lunch (with
Warner wines) is served daily.

Warner owns 350 acres of vineyards. Although none are planted on the
winery property, all are within a thirty-three mile radius of Paw Paw. These
vineyards are planted to twelve different grape varieties, largely French-
American hybrids. Warner processes three-million gallons of juice annually; its product line contains forty-five labels—table wine, sparkling
wine, fortified wine, and fruit wine—as well as twelve different fruit juices
ranging from apple to fruit punch. Capriccio Spumante, Warner Brut,
and Cranberry wine (a blend of white table wine and cranberry juice)
are among Warner's most popular products.

Open all year, except Thanksgiving, Christmas, New Year's, and Easter. Sum-
mer: Monday-Saturday, 9-6; Sunday, 12-6. Winter: Monday-Saturday, 10-5;
Sunday, 12-5. Guided tours. Free tasting. Retail outlet: wine, gifts, glassware.
Grape and Wine Festival (September). Wine Garden Restaurant open daily,
11-4. Access for the handicapped limited to tasting room. Directions: Interstate
94 to Paw Paw (exit 60). At exit, go two blocks north to winery.

Traverse City

Chateau Grand Traverse
12239 Center Road
Old Mission Peninsula
Traverse City, Michigan 49684
Telephone: (616) 223-7355
Owner: Edward O'Keefe

In view of Michigan's bitter winters, when temperatures have been known to drop to thirty-eight degrees below zero, the very existence of Chateau Grand Traverse's all-vinifera vineyard is no mean achievement. It is, in fact, the only such vineyard in the state. By virtue of its location on Old Mission Peninsula, where climate is moderated by the waters of Grand Traverse Bay, these delicate premium grape varieties have managed not only to survive, but to thrive, producing prize-winning varietal wines.

Founded in 1974 by wine importer Edward O'Keefe, Chateau Grand Traverse's model winery and its surrounding vineyard are located on a hill north of Traverse City. The original vineyard was planted by German viticulturists largely to Johannisberg Riesling and Chardonnay, with smaller amounts of Merlot; it is currently being expanded by another fifty acres. From these grapes, its German-trained winemaker, Roland Pfleger, produces 7,500 cases of estate-grown-and-bottled wine annually, including the varietals Johannisberg Riesling (dry, semi-dry, late harvest, and ice wine), Chardonnay, and Merlot natural. The winery is particularly known for its Riesling and Chardonnay, both of which are national award winners.

Guided tours of Chateau Grand Traverse are offered in spring, summer, and fall. These include a look at the vineyard and working winery and conclude with a tasting. The winery also sponsors a Spring Wine Festival (Memorial Day weekend), featuring jazz and blues bands and ethnic foods, as well as a Fall Winefest (Labor Day weekend) with German music and a 15-kilometer vineyard run. There are picnic facilities on the grounds.

Open all year: Monday-Saturday, 11-6; Sunday, 12-6. Guided tours: spring, summer, and fall. Retail outlet: wine, gifts. Seasonal festivals. Picnic facilities. Access for the handicapped. Directions: From Traverse City, west on US 31 to M-37. North on 37 for 7 miles to winery.

MINNESOTA

Hastings

Alexis Bailly Vineyard
18200 Kirby Avenue
Hastings, Minnesota 55033
Telephone: (612) 437-1413
Owner: The David Bailly Family

Family-owned-and-operated Alexis Bailly Vineyard was founded in 1976 by Minneapolis attorney David Bailly and his six children. Located on the outskirts of the town of Hastings (fifteen miles southeast of Minneapolis), the winery is housed in a log cabin built expressly for its current use. A tour reveals the Baillys' predilection for using native materials—especially wood. Fermentation tanks are made of redwood; the wine is aged in oak barrels; and the building itself is constructed of Minnesota White Pine and local flagstone.

A sign near the vineyard reads, "Where the grapes suffer," attesting to the harsh Minnesota climate: the vines have to be removed from their trellises and mulched with straw or covered with soil to see them through the winter. Bailly's ten-acre vineyard is planted largely to French-American hybrids and a smaller quantity of Swenson Red. From these grapes and additional grapes and fruit purchased from local growers, the winery produces 5,000 gallons of wine yearly; the varietals Seyval Blanc, Maréchal Foch, and Léon Millot; a mead derived from Minnesota honey; and two fruit wines—apple and an apple/grape blend. Bailly is best known for its Léon Millot.

The winery is named for David Bailly's great-great-great-grandfather, a colorful character who was run out of the territory for selling whiskey to the Indians in the early 1800s, but who later returned to found Hastings in 1854.

Open May-November: Tuesday-Saturday, 10-5. Guided tours by appointment. Free tasting. Retail outlet: wine. Picnic facilities. Directions: From Hastings, south on Highway 61 to 180th Street. 180th Street west one mile to winery.

Stillwater

Northern Vineyard Winery
402 North Main Street
Stillwater, Minnesota 55082
Telephone: (612) 430-1032
Owners: Minnesota Winegrowers Cooperative, Inc.

An old mill complex on the Croix River houses the winemaking facilities, tasting room, and retail outlet of Northern Vineyards Winery, a cooperative venture of eleven local grapegrowers formed in 1983. The winery's first vintages ('83 and '84) were produced at David Macgregor's Lake Sylvia Vineyards Winery in Maple Lake. At its new location in Stillwater, the restored Staples sawmill, David Macgregor continues as winemaker. Production for Northern Vineyards Winery is currently 3,000 gallons per annum; grapes are supplied by the cooperative's vineyards, which total about thirty acres in area.

Planted in these vineyards (located in the south-central part of the state) are French-American hybrids such as Maréchal Foch, de Chaunac, Seyval Blanc, Aurora, and improved varieties developed by Wisconsin hor-

ticulturist Elmer Swenson to withstand the state's cold winters: Edelweiss, Kay Gray, St. Croix, and LaCrosse. From these grapes, the winery produces predominantly white and rosé wines, with a limited amount of red. The whites are cool-fermented; the reds, aged in white oak. Wines produced are frequently blends, such as the 1983 White Table Wine made from Aurora, LaCrosse, Edelweiss, and Elvira grapes. These vintage-dated white blends are the winery's most popular. In addition, Northern Vineyards produces a red table wine and three vintage-dated varietal wines: Seyval Blanc, Maréchal Foch Rosé, and de Chaunac.

With three days' notice, brief guided tours of the winery are available. Tastings are available for a small charge anytime the winery is open. (This charge is refunded if wine is purchased.)

Open all year: Thursday-Sunday, 11-6. Guided tours and tasting as above. Retail outlet: wine. Directions: From Minneapolis-St. Paul take Interstate 94 east to Highway 95. North on 95 to Stillwater. Winery located in Staples Mill complex, three blocks north of the bridge on 95 (called Main Street locally).

MISSOURI

Augusta

Mount Pleasant Vineyards
101 Webster Street
Augusta, Missouri 63332
Telephone: (314) 228-4419
Owners: Lucian W. and Eva B. Dressel

Mount Pleasant Vineyards sits on a high bluff overlooking the bottom land of the Missouri River on the edge of the town of Augusta, once a lively center of winemaking and grape growing. The winery was originally founded in 1881 by Lutheran minister and vintner Friedrich Münch, who in addition to making highly regarded wines, wrote *School for American Grape Culture*, an early treatise on viticulture.

Mount Pleasant was one of eleven wineries established by Augusta's German settlers. All of the facilities were closed by Prohibition. In 1968, after years of neglect, Mount Pleasant was purchased by accountant and wine lover Lucian Dressel and his wife, Eva. Today the winery's three underground cellars are again filled with casks of aging wine. Its high-ceilinged one-story brick building again houses the fermentation, pro-duction, and bottling activities of the winery, as well as providing space for a tasting room.

Mount Pleasant's forty acres of vineyard are planted primarily to French-American hybrids. Twenty-five different varieties of grapes are grown, and Mount Pleasant produces between fourteen and twenty labels; its Seyval Blanc has been an award winner.

Guided tours of the winery are given on the hour and take about twen-ty minutes. These cover its vineyard, production facility, and cellars. After-wards, the whole Mount Pleasant line, even its sparkling wines, can be tasted at no charge.

There is picnicking on the terrace to the winery. From the shop across the way—which is owned by Eva Dressel—you can purchase cheese and sausage to accompany a wine of your choice.

Open all year, except Thanksgiving, Christmas, and New Year's: Monday-Saturday, 10-5:30; Sunday, 12-5:30. Guided tours. Free tasting. Retail outlet: wine. Festivals: Strawberry Festival (May), Harvest Weekend (Fall). Sausage and cheese shop. Picnic facilities. Access for the handicapped only to the tasting room and terrace. Directions: From St. Louis, Highway 61 west to Route 94. 94 southwest to Augusta.

Benton

Moore-Dupont Wineries
Interstate 55 (P.O. Box 211)
Benton, Missouri 63736
Telephone: (314) 545-4141
President: Jean René Dupont

Missouri's newest winery, Moore-Dupont, is associated with the state's largest vineyard holdings—170 acres planted to French-American hybrids, native American varieties, and viniferas. It was founded by surgeon Jean René Dupont and farmer Handy Moore in 1977. The location they chose, in the Mississippi River Valley, is a new area for Missouri winegrowers, one which has been compared in climate with Alsace-Lorraine. Dupont and Moore brought in California-trained Bill Lamberton to serve as winemaker.

Moore-Dupont's modern facility in Benton currently produces the com-pany's table wines and juices. A second winery is scheduled to open in Springfield to produce its champagne and sparkling wines. Current production is 40,000 gallons annually, largely still table wines with a few *methode champenoise* sparkling wines and fruit wines. Wines include the varietals Villard Blanc, Seyval Blanc, Riesling, Catawba, and Chambourcin; a generic chablis; and a mead. Moore-Dupont is particularly known for its Villard Blanc. Tours of the winery include a look at its cellar, lab, and bottling line and conclude with a tasting of the Moore-Dupont product. Usually six wines are opened each day. There is a canopied picnic area adjoining the winery; you can purchase lunch inside at a convenient delicatessen.

Open all year. Summer: daily, 10-8. Winter: limited hours (check ahead). Guided tours (30 minutes). Free tasting. Retail outlet: wine, gifts. Festivals: Second weekend in May and October. Delicatessen. Picnic facilities. Access for the handicapped. Directions: From St. Louis, take Interstate 55 south to Benton exit. Winery at exit.

Hermann

Hermannhof Vineyards
330 East First Street
Hermann, Missouri 65041
Telephone: (314) 486-5959
Owner: James F. Dierberg

Although Hermannhof Vineyards was founded in 1978, its roots, like those of its neighbor, Stone Hill (see entry following), are in the nineteenth century. In the 1850s, Hermannhof's multistory brick building served as a brewery. Restored by banker James F. Dierberg, the winery, which has ten historic stone cellars, now produces between 12,000 and 15,000 gallons of wine annually.

Hermannhof's forty-acre vineyard, located ten miles away, is planted to Seyval Blanc, Vidal Blanc, Villard, Norton, Chambourcin, Delaware, Vignoles, and Cayuga. From these grapes and additional purchased crops the winery makes still table wine and *methode champenoise* sparkling wine. Varietal wines include Norton, Chambourcin, and Seyval Blanc. The winery is best known for the latter. Its proprietary blends, White Lady and Settler's Pride, are also exceptional.

Guided tours include its landmark cellars and the Smoke Haus, where sausages and cheeses are processed. The wine tasting takes place in Hermannhof's deli/bar, or, in fine weather, in the adjacent courtyard. Usually ten wines are available for tasting; you can purchase homemade and imported cheeses and sausage to nibble with your wine. A craft center is located in the old cooperage, where local artisans exhibit their wares.

Open all year: Monday-Saturday, 9:30-5:30; Sunday, 12-5:30. Guided tours available March-December, daily 10-4 (weekends only January and February). Free tasting (minimal charge for premium wines). Retail outlet: wine, gifts. Crafts shop. Maifest (third weekend in May); Oktoberfest (every weekend in October). Smoke Haus: smoked sausage and cheeses. Picnicking facilities. Limited access for the handicapped. Directions: Interstate 70 to Highway 19 south. At Hermann, east on Highway 100 (First Street) to winery.

Stone Hill Wine Company
Highway 19 (Route 1, Box 26)
Hermann, Missouri 65041
Telephone: (314) 486-2221
Owners: L. James Held and Betty A. Held

Stone Hill Winery is perched dramatically on a hill overlooking the picturesque town of Hermann and the Missouri River. It was founded in 1847 by German immigrant Michael Poeschel and by 1900 was the third largest winery in the world (second in the nation), with an annual output of 1¼ million gallons. Here, on the sheltered south bank of the river, Poeschel built a formal three-story brick building with shuttered windows and a white cupola. The first floor of the building, originally the winery office, now serves as its tasting room/retail area. The living quarters above are preserved as they were, and the third floor contains an attic museum of wine artifacts. The cupola, reached by stairs from the third floor, offers sweeping views of the Missouri River and of Hermann's German-inspired architecture.

Prohibition and the Depression brought an end to Stone Hill's—and Missouri's—preeminent viticultural position. And for forty years thereafter, mushrooms were cultivated in Stone Hill's damp limestone cellars. James and Betty Ann Held, local grape growers, bought the property in 1965 and have been returning the wonderful assemblage of landmark buildings and caves to their original use ever since. The surrounding hills, where the original vineyards were located, are once again growing grapes. The Helds have planted sixty acres to Seyval Blanc, Vidal Blanc, Vignoles, Chancellor, Norton, Missouri Riesling, Niagara, Colobel, and Villard Noir. From these crops and additional purchased grapes (225 tons), they produce 34,000 cases of wine annually. Still table wine accounts for the major part of Stone Hill's product line; smaller amounts of sparkling and fruit wines are offered as well.

Extending down the hill are the ruins of the original winery. Underneath its stone arches and brick stacks lie the famous Stone Hill Cellars, which required twenty years to dig and which are said to be the most extensive in America. In this underground labyrinth, rows of stainless-steel fermentation tanks and aging barrels have replaced the mushrooms which sustained Stone Hill in the '20s and '30s. Next to the winery, a red horse barn has been converted into a restaurant serving German cuisine.

Guided tours of Stone Hill cover the history of Hermann and the winery, and include a look at the underground cellars. (A $1 charge for tours is used for the restoration and maintenance of the buildings.) Tasting is available at no charge with or without the tour. There are picnicking facilities on the grounds, and popular festivals are held during the year, including the Maifest (third weekend in May) and the Octoberfest (every weekend in October).

Stone Hill's most popular wine is its vintage-dated Norton. (Grapes for this wine, incidentally, were cultivated from cuttings from the half-acre vineyard remaining on the estate when the Helds purchased it.)

Open daily all year, except Thanksgiving and Christmas. Summer months: Monday-Saturday, 8:30-7:30; Winter, Monday - Saturday, 8:30-5; Sundays, all year, 12-6. Guided tours, $1. Free tasting (grape juice for children). Retail outlet: wine, cheese, gifts. Museum. Festivals: Maifest (third weekend in May), Octoberfest (every weekend in October). Restaurant: Vintage 1847, open daily all year, 11-10 (German-style cuisine). Picnicking on patios. Directions: From Interstate 70, 15 miles south on Highway 19. (Winery on southern boundary of Hermann).

St. James

Ferrigno Vineyards and Winery
Highway B (Route 2, Box 227)
St. James, Missouri 65559
Telephone: (314) 265-7742
Owners: Richard and Susan Ferrigno

Of all the wineries located in and around St. James, Ferrigno has one of the prettiest settings and is one of the few with vineyards contiguous to its winery. Ferrigno's working winery and tasting room are housed in a renovated vintage-1930 dairy barn, behind which a lovely shaded garden faces the vineyard. The working winery is located on the ground floor of the barn; the tasting/retail area is in the loft.

Tours of the working area are by appointment only, except on festival weekends (twice yearly) when they are regularly scheduled. These festivals, called Wine Expos, are held jointly with the other area wineries in the months of June and September.

Ferrigno's vineyards were established by Italian immigrants who settled the area in the 1920s. In addition to these venerable vines, some of which are still producing, the Ferrignos planted an additional twelve acres to French-American hybrids when they purchased the property in 1976.

Varieties grown include Seyval Blanc, Vidal Blanc, de Chaunac, Chelois, Baco Noir, and Concord. Ferrigno's first crush was completed in 1981 (all grapes are hand-harvested). White wines are cold-fermented; the Chelois is aged for two years in Missouri white oak. The winery produces only still table wine, and its current output is 2,000 gallons yearly. Ferrigno is best known for its varietals Chelois and Seyval Blanc, and its proprietary blend, Vino de Famiglia.

There is no charge for the regular tasting, but for $5 per person you can enjoy a special, relaxed tasting in the garden, with a lunch of bread, cheese, and sausage to accompany the Ferrigno product.

Open all year, except major holidays: Monday-Saturday, 10-6; Sunday, 12-6. Tours by appointment only. Free tasting. Special tasting as above. Retail outlet: wine, gifts, grape juice, cheese. Wine Expos: second weekend in June, third weekend in September. Garden for picnicking. Access for the handicapped to garden only. Directions: Interstate 44 to St. James turn-off. North access Road B east to Winery (beyond St. James Winery on your left).

Rosati Winery
Highway KK (Route 1, Box 55)
St. James, Missouri 65559
Telephone: (314) 265-8629
Owner: Robert H. Ashby

Rosati's wines reflect the Italian heritage of the early settlers of this region, known broadly as the Ozark foothills. Established in 1934, the winery is located between Cuba and St. James in a vintage brick building set back from the access road (KK) which parallels Interstate 44. Although there are no vineyards on the property (they are located northeast of the winery), there is an attractive patio off the tasting room for picnicking and a pond out front, bucolic touches sure to please the visitor. Tours of the winery are self-guided; the route is well marked with pictures and signposts explaining its history and ethnic style of winemaking, its modern winemaking equipment, and its fifty-year-old cellars.

Afterwards you can sample Rosati's whole product line, over twenty differents labels, including varietals, proprietary blends, sparkling wines, and fruit wines. The best known is its old-fashioned Concord, a sweet red Italian-style wine.

Rosati's twenty-acre vineyard is planted to Elvira and Concord. From these and additional purchased grapes, the winery produces between 15,000 and 20,000 gallons annually.

Open all year: Monday-Saturday, 8-6; Sunday, 12-6. Self-guided tours. Free tasting. Retail outlet: wine, gifts. Wine Expos: second weekend in June, third weekend in September. Picnicking facilities. Access for the handicapped limited to main winery areas. Directions: From St. Louis, take Interstate 44 to exit F and ZZ. Cross overpass and continue west on access road ZZ (it becomes KK) to winery.

St. James Winery
540 Sidney Street
St. James, Missouri 65559
Telephone: (314) 265-7912
Owners: James and Patricia Hofherr

St. James Winery, founded by James and Patricia Hofherr in 1970, is one of the largest wineries in the Ozark Highlands, with an annual production of between 30,000 and 40,000 gallons. The winery, with its stone and cedar facade and red roofed, gabled tasting room, sits on land that was formerly owned by the St. James Aircraft Company. (The fledgling airport earned a place in history on June 25, 1926, when Charles Lindbergh landed there and offered local residents airplane rides for a small fee.)

Tours of the working winery are self-guided; informative signs posted along the way explain the press, crusher-stemmer, filters, tanks, bottling machinery, and other steps in the winemaking process. In the tasting room/retail area all the available St. James wines can be sampled at no charge. The product line, over twenty strong, encompasses sweet, semi-dry, and dry table wines, sparkling wines, fruit wines, and mead. Next to the tasting room are picnic tables sheltered by a grape arbor. (Cheese and crackers can be purchased in the tasting room if you didn't pack lunch.)

St. James owns a seventy-acre vineyard planted to Concord, Catawba, Niagara, Villard Noir, Rougeon, Munson, Cynthiana, Isabella, Vidal Blanc, Seyval Blanc, and Chancellor. The vineyard, located two miles east of the winery, can be toured by appointment.

Open all year: Monday-Saturday, 8-6; Sunday, 12-6. Self-guided tours, daily; guided, by appointment. Free tasting. Retail outlet: wine, wine-related merchandise, gifts, cheese. Ozark Highland Wineries Expos: second week in June, third week in September. Picnic facilities. Access for the handicapped. Directions: Interstate 44 to St. James turn-off. North Access Road B, east to winery.

OHIO

Cleveland Heights

Cedar Hill Wine Company
2195 Lee Road
Cleveland Heights, Ohio 44118
Telephone: (216) 321-9511
Owner: Dr. Thomas W. Wykoff

Winemaking operation of surgeon Thomas W. Wykoff, head of Ear, Nose, and Throat Department of Cleveland's St. Luke's Hospital. Dr. Wykoff also owns the Au Provence Restaurant, upstairs. Winery, bonded in 1974, produces first-rate wines (2,000 cases annually) from purchased grapes (Lake Erie district). Wines marketed under Chateau Lagniappe label include vintage-dated varietals and generics: Seyval Blanc, Vidal Blanc, Chardonnay, Chambourcin, Dutchess, Terminal Red, and Modern Dance Rosé. Best known for Chardonnay, Seyval Blanc, and Chambourcin. Open all year: Monday-Saturday, 11-6. Guided tours, 15-30 minutes. Free tasting. Retail outlet: wine, gifts. Restaurant: same hours (reservations required on Saturday). (You can easily visit the small cellar for a tasting before you sit down for dinner.) Access for the handicapped. Directions: Interstate 271 to Cedar exit. Cedar Road west to Lee Road (10 minutes). Restaurant and winery on east side of Lee.

Conneaut

Markko Vineyard
South Ridge Road (R.D. 2)
Conneaut, Ohio 44030
Telephone: (216) 593-3197
Owners: Arnulf Esterer and Thomas H. Hubbard

Markko Vineyard, a small premium winery nestled in the woods three miles from the shore of Lake Erie in northeastern Ohio, was established in 1968 by industrial engineer Arnulf Esterer, one of the first growers to introduce the vinifera grape to Ohio, an achievement he accomplished after a two-year apprenticeship with noted wine authority Dr. Konstantin Frank. In 1969, Esterer began planting his fourteen-acre vineyard to Chardonnay, Johannisberg Riesling, Cabernet Sauvignon, and Cham-

bourcin. He is the winemaker; his wife, Kate, a rural mail carrier, and their four children all lend a hand in the daily operation of the winery and management of the vineyard.

Markko produces 5,000 gallons of wine annually. All wines are vintage-dated and are produced from the firm's own grapes. Its product line includes the varietals Chardonnay, Johannisberg Riesling, and Cabernet Sauvignon, and proprietary red and white blends under its Covered Bridge and Underridge labels. Markko is especially known for its vintage varietals.

Guided tours of the winery are usually given Monday through Saturday, but a call ahead is advised. The tour includes a look at both the vineyard and winery and concludes with a tasting, for which there is a 50-cent charge per variety. (This is refunded with the purchase of a case of Markko wine.)

There are picnic facilities on the grounds; the winery also capitalizes on its scenic Ashtabula County location by sponsoring a covered bridge bicycle tour and picnic every spring. The event includes a barbecue at the winery after the fifteen-mile tour, which passes over three covered bridges, is completed. (There is an optional two-mile walking route as well.)

Open all year: Monday-Saturday, 11-6. Guided tours, appointments advised. Tasting: $.50 per wine. Retail outlet: wine. Special events: Markko Jug Day (June), Covered Bridge Bicycle Tour and Picnic (June). Picnic facilities. Limited access for the handicapped. Directions: From Cleveland, Interstate 90 to Kingsville exit. North and west on South Ridge Road to winery.

Madison

Chalet Debonné Vineyard
7743 Doty Road
Madison, Ohio 44057
Telephone: (216) 466-3485
Owners: The Debevc Family

Chalet Debonné, located on the south side of Interstate 90 about twenty-five miles west of Markko Vineyard, is headed by father and son Tony J. and Anthony P. Debevc. Tony is viticulturist; his son holds a pomology degree from Ohio State University.

The two built Chalet Debonné's two-story winery next to the Concord vineyard planted by Anthony's grandfather before Prohibition. The winery's working elements are located in an underground cellar; above are the hospitality and retail areas, dominated by a large fireplace and finished with weathered barn siding. A greenhouse addition offers a good view of the vineyards beyond.

Chalet Debonné, established in 1970, now has fifty acres planted to French-American hybrids and viniferas. From this crop and additional purchased grapes, the winery produces 36,000 gallons of table wine annually, including vintage-dated varietal wines: Cabernet Sauvignon, Chardonnay, Seyval Blanc, Vignoles, and Villard Blanc. Chalet Debonné is also highly regarded for its red and white Labruscas.

Tours of the winery are available on the hour and are guided by family members. Free tastings are offered, or for a small sum you can purchase a tasting tray which includes cheese, crackers, and sausage.

Open February-December: Tuesday, Thursday, Saturday, 1-8; Wednesday and Friday, 1-11. Open January for sales only: Saturday, 1-4; other days by appointment. Guided tours on the hour. Free tasting, or tasting trays with cheese and sausage ($2-$2.50). Retail outlet: wine. Limited access for the handicapped. Directions: From Cleveland, Interstate 90 to Madison exit. South on 528 (over Grand River) to Griswold Road (first left). Go east on Griswold to Emerson. Take Emerson north to Doty, and Doty east to winery. There are signs marking the way from intersection of Griswold and 528.

Grand River Vineyard
5750 Madison Road
Madison, Ohio 44057
Telephone: (216) 298-9838
President: Willett Worthy

Established in 1971 by ex-banker Willett Worthy, Grand River Vineyard, like its neighbor Chalet Debonné, is a family-run winery. Worthy's wife, Carroll, and children, Adrienne and John, pitch in when needed. Worthy is one of the most successful growers of vinifera in the Midwest. His rustic L-shaped winery is in a wooded area next to the vineyard, which is planted to thirteen different grape varieties, including viniferas and French-American hybrids. From these grapes and additional purchased fruit, Grand River produces about 9,000 gallons of wine annually. It is noted for its varietals, particularly the dry reds, such as Merlot, and German-style whites, including Gewürztraminer.

   Guided tours of the winery include the working portion of the open-air pavilion, with its crush pad and presses, as well as indoor production facilities. Grand River's small tasting room is furnished with an antique bar; a second, larger room with beamed ceiling, fireplace, and French doors leading to the pavilion, is frequently reserved for private parties. The pavilion is also used for picnicking and for community theatre productions sponsored by the winery in March, June, August, and November. Performances are held inside in the winter months.

Open all year: Monday and Thursday, 1-8; Wednesday, 1-10; Friday and Saturday, 1-6. Guided tours. Free tasting. Retail outlet: wine, wine accessories, gifts. Community theatre productions as above. Picnic facilities. Limited access for the handicapped. Directions: Interstate 90 to Madison-Thompson exit (State Road 528). Take 528 south for 5 miles (towards Thompson) to winery.

Manchester

Moyer Vineyards
3859 U.S. Highway 52
Manchester, Ohio 45144
Telephone: (513) 549-2957
Owners: Kenneth L. and Mary B. Moyer

When you plan an outing to Moyer Vineyards, you have a choice of transport: either follow Highway 52 along the scenic Ohio River, or come by boat and dock at the winery landing. Established in 1972 by Kenneth and Mary Moyer, Moyer Vineyards produces 10,000 gallons of wine annually from its ten-acre vineyard which is planted to Chambourcin, Vidal Blanc, and Villard Blanc; additional grapes are purchased as needed. Moyer's product line includes varietal table wines and *methode champenoise* sparkling wine (produced in its New Braunfels, Texas, facility).

Fifteen-minute tours of the winemaking operation are offered. You might like to follow the tour and tasting with a meal at the winery's upstairs restaurant (Mary Moyer presides over the kitchen). In nice weather, you can eat *al fresco* on the deck overlooking both the vineyard and the river.

Open all year: Monday-Thursday, 11:30-9; Friday and Saturday, 11:30-10. Guided tours: 11:30-4, or by appointment. Free tasting. Retail outlet: wine. Restaurant: serves lunch and dinner. (Same hours as winery.) Limited access for the handicapped. Directions: From Cincinnati, take U.S. 52 70 miles east to Manchester.

Middle Bass Island

Lonz Winery
1 Fox Road
Middle Bass Island, Ohio 43446
Telephone: (419) 285-5411
Owner: Paramount Distillers Inc.

Middle Bass is one of a cluster of three Ohio islands often referred to as the wine islands; the others are Put-in-Bay and Isle St. George. Lonz

Winery, located on Middle Bass, has a venerable history; it was established as the Golden Eagle Winery during the Civil War. By 1875, it had become one of the largest wine producers in the United States. In 1884, Peter Lonz started producing wines on Middle Bass and it was his son, George, a colorful figure, who designed the substantial "castle," with stone battlements and turrets, that houses the winery to this day.

Tours begin with a short slide presentation covering the winery's history; guides then lead you through its 100-year-old cellars, cut deep into the native limestone. At the tour's conclusion, wine appreciation is discussed while you sample the house product.

Lonz is just resuming operation after ten years of inactivity. Its ten-acre vineyard is being replanted to Catawba, Chardonnay, Gewürztraminer, and Riesling. Current production is limited to *methode champenoise* sparkling wine and oak-aged sherry.

This island winery is a good place to spend the day; you can reach Middle Bass Island by private boat, public ferry, or even by plane. Lonz has its own marina (capacity 150 boats). You can purchase snacks, pizza, and sandwiches on the premises; a terrace for picnicking overlooks the lake marina. The island offers facilities for swimmers as well.

Open May-September: Monday-Thursday, 12-8; Friday, 12-10; Saturday, 12-12; Sunday, 1-6. Open October: weekends only. Guided tours every half hour beginning at 12:30 daily (last tour 4:30): adults, $1.25; children, 50¢. Tasting included in tour charge. Retail outlet: wine, nautical gifts. Grapestomping festival (third weekend in September). Restaurant: sandwiches, pizza, and light snacks; open daily 12 to closing. Picnic facilities. Access for the handicapped to all areas except cellar. Directions: Ferry from Catawba Point (Miller Ferry), Port Clinton (Parker Ferry), or Put-in-Bay Water Taxi. Air service: Island Airlines, Port Clinton.

Morrow

Valley Vineyards Farm
2041 East U.S. 22/Ohio Route 3
Morrow, Ohio 45152
Telephone: (513) 899-2485
Owners: Ken and James Schuchter

Established in 1969. Winery located on bank of Little Miami River in the hilly countryside of southern Ohio. Building—faced with new brick, stucco, and stained timbers—incorporates a large old barn. Valley Vineyards is one of the larger growers in Ohio (owners were farmers before they were vintners). Forty-five acre vineyard planted to French-American hybrids and native American varieties. Production: 30,000 gallons annually (purchased grapes supplement the harvest). Labels include varietals (especially de Chaunac), proprietary wines, and generics: Hillside Red, Chablis, Sauterne, and Sangria. Open all year, except Christmas: Monday-Thursday, 11-8; Friday and Saturday, 11-11. Self-guided tours (guided tours by appointment): 11-8. Tastings: 10¢ per ounce. Retail outlet: wine, gifts. Valley Vineyards Wine Festival (last weekend in September). Restaurant: serves dinner on alternate Fridays only. Picnic facilities. Access for the handicapped. Directions: From Cincinnati, Interstate 71 northeast to exit 48 (Kings Island interchange). South on 48 3 miles to Route 22, then east on 22 3 miles to winery.

Port Clinton

Mon Ami Wine Company
3845 East Wine Cellar Road
Port Clinton, Ohio 43452
Telephone: (419) 797-4445
Owner: Paramount Distillers Inc.

Before or after your tour of the island wineries of Lonz and Heineman, be sure to make time to visit Mon Ami's historic winery, located near the ferry slip in Port Clinton. Mon Ami has been in operation since 1870. Its century-old stone winery building is handsomely situated in a wooded setting.

There is a nominal charge for tours which cover the lower level of the winery building, including the fermenting room where Mon Ami's *methode champenoise* sparkling wine is aged *en tirage*, the riddling room,

and the aging and storage areas. Afterward you can visit the winery's combined tasting bar/retail area, enjoy a meal in one of its two restaurants, or picnic in the shaded outdoor dining area. The Chalet, the newer and more informal of the two restaurants, serves sandwiches and ribs. Depending on the weather, you can eat outdoors by the fountain, or inside by the fireplace. Mon Ami Restaurant, with its more extensive menu, is one of the most popular in the area.

Mon Ami purchases all of the grapes needed to produce 50,000 gallons of wine annually. Seventy percent of its line is still table wine; twenty percent, sparkling wine; and the balance, fortified wine. The best known are its *methode champenoise* sparkling wines and native American blends.

Open all year. Guided tours (adults $1; tour including tasting with cheese and crackers, $1.50): May 1-October 1, daily, 2-4; winter weekends, tours by request. Tasting with or without tour. Retail outlet: wine, gifts. Chalet Restaurant and Mon Ami Restaurant, open daily. Access for the handicapped. Directions: Ohio Turnpike to exit 6 (Route 53). Route 53 north to Port Clinton area and Route 2. East on Route 2 to winery.

Put-In-Bay

Heineman Winery
Catawba and Cherry Avenue (P.O. Box 300)
Put-in-Bay, Ohio 43456
Telephone: (419) 285-2811
Owner: Heineman Beverage, Inc.

Heineman Winery is on the wine island of South Bass, only a short ride via water taxi from Middle Bass Island's Lonz winery. (South Bass Island is also called Put-in-Bay after the principal village located thereon.) The only winery remaining on the island (although there are other vineyards), it is now operated by Louis Heineman, but was established in 1887, when Louis's grandfather, Gustav, planted a vineyard. Two years later, when excavating a site for the winery, he discovered an unusual cave of spectacular bluish-white celestite crystals, some as large as three feet long and eighteen inches across. Said to be the only cave of this formation in the United States, it became a great tourist attraction.

Tours of Heineman are offered from May 15 to September 15. Tours include the whole working facility, the cave, and a glass of wine or grape juice. There are picnic tables on the front lawn and a lovely wine garden behind the winery.

Heineman's vineyards, totaling 35 acres, are scattered about the island and are planted to Catawba, Niagara, Ives, Delaware, and Concord. The winery produces 30,000 gallons of still table wine annually and is noted for its Pink Catawba and Island Chablis.

Open daily all year: October 1-May 1, 9-5; June 1-September 15, 9-10. Guided tours (include cave and glass of wine or juice): May 15-September 15, 11-5 (adults, $2; children, $1). Retail outlet: wine. Wine garden: cheese and wine. Picnic tables. Limited access for the handicapped. Directions: Ohio Turnpike to exit 6 (Route 53). Route 53 north to Catawba Point (Miller Boat Lines). From Middle Bass Island: water taxi to Put-in-Bay.

Sandusky

Mantey Vineyards
917 Bardshar Road
Sandusky, Ohio 44870
Telephone (419) 625-5474
Owner: Paramount Distillers Inc.

One of state's leading producers of wine made exclusively from Ohio-grown grapes. Originally established in 1880 in Venice as a small fruit farm. Less than a half-hour drive east of Paramount Distillers' other mainland holding, Mon Ami. (Paramount also owns Lonz on nearby Middle Bass Island.) White barn winery incorporates tasting and retail area. Guided tours on the hour cover both vineyard and winery. Free tasting of four Mantey wines daily (grape juice also served). Thirty-acre vineyard behind winery planted to Labrusca and French-American hybrids. 50,000 gallons or more produced annually. Additional grapes purchased from local growers. Product list in-

cludes table, sparkling, fortified, and fruit wines. Most noted for French-American and Labrusca varietals and altar wines. Open all year: daily, 9-5. Tours: June 1-September 1, Monday-Saturday, 10-3; other months by appointment only. Retail outlet: wine, gifts, home winemaking supplies. Directions: West of Sandusky via Route 2 or 6. Winery at junction of two routes. (The winery's eight large juice storage tanks are visible from the road.)

Silverton

Meier's Wine Cellars
6955 Plainfield Pike
Silverton, Ohio 45236
Telephone: (513) 891-2900
Owner: Paramount Distillers Inc.

Oldest and largest winery in Ohio. Current capacity: 2½ million gallons. Located ten miles north of downtown Cincinnati. Three-hundred-acre vineyard on Isle of St. George planted to vinifera, French-American hybrids, and Labruscas. Excellent tour covers crushing, pressing, aging, packaging, and tasting, with commentary on history of wines sampled and remarks on their storage and service. Film and exhibit of wine artifacts included. Product list includes varietals and generics. Best known for #44 Cream Sherry and Haut Sauterne. Open all year, except Thanksgiving, Christmas, and New Year's: Monday-Saturday, 9-5. Guided tours: Memorial Day-October 31, Monday-Saturday, on the hour between 10 and 3. Tours other months by appointment. Tasting (nominal charge). Retail outlet: wine, gifts. Snack bar with wine and cheese service in formal garden. Limited access for the handicapped. Directions: From Cincinnati, Interstate 71 to exit 12 (Montgomery Road). West on Montgomery Road to Plainfield Pike. North on Plainfield Pike to winery (on your left).

Troy

Stillwater Wineries
2311 State Route 55 West
Troy, Ohio 45373
Telephone: (513) 339-8346
Owners: Allen Jones and James Pour

Stillwater Wineries, located north of Cincinnati in the greater Miami River Valley, is probably the only winery in the state with a picnic area on its roof. (Opened in 1982, it is carved into the side of a hill. And with the exception of its glass front, everything—lab, offices, tasting room, aging cellar, storage, etc.—is completely covered by earth and landscaping.)

Stillwater's twenty-two-acre vineyard, four miles away, is planted to French-American hybrids and Labruscas. (Owners Allen Jones and James Pour began as grape growers; the winery was a natural extension of their harvests.) Stillwater produces as much as 17,000 gallons of wine yearly, or as little as 3,000, depending on the severity of the winter. Still wines account for ninety-six percent of its production; the balance is *methode champenoise* sparkling wine and fruit wine. The list of fifteen labels is devoted largely to varietal wines; two blends round out the selection. Cayuga White is considered the best wine; the most popular is a blend, Stillwater White.

Tours of the winery are given regularly from New Year's Day through Thanksgiving. These cover the whole winemaking process and take about half an hour, depending on the curiosity of the participants. Afterwards you can taste Stillwater's wines in its large tasting room, which contains seating for 120 people. Snacks, cheese, and homemade bread are also sold. And you are most welcome to bring a picnic for either lunch or dinner, since the winery is open until eleven o'clock every night.

Open all year. New Year's Day-Thanksgiving: Tuesday-Friday, 4:30-11; Saturday, 1-11. Thanksgiving-December 31: Monday-Saturday, 1-11. Guided tours: Friday and Saturday evenings after 7. Other days and times by appointment. Tasting: $1.50 per tray of six wines. Retail outlet: wine. Picnic facilities. Greater Miami Valley Grape Stomp and Goat Races (first Saturday after

Labor Day.) Limited access for the handicapped. Directions: From Cincinnati, north on Interstate 75 to State Route 55. Route 55 west ½ mile to winery.

West Milton

Heritage Vineyards
6020 South Wheelock Road
West Milton, Ohio 45385
Telephone: (513) 698-5369
Owners: Edward Stefanko and John Feltz

Located approximately ten miles southwest of Stillwater Wineries in remodeled 100-year-old barn. Cedar-paneled tasting room offers view of vineyards from windows of second-story loft. Twenty-acre predominantly French-American hybrid vineyard established in 1972; winery opened in 1978. 15,000-gallon capacity. Product line primarily varietal table wines; also offers both methode champenoise *sparkling wine and Heritage Vineyards Ohio Champagne. Open all year. June-October: Tuesday-Saturday, 1-10; December-April: Friday and Saturday, 1-10. Tours on request. Tasting trays: $2.50 (six wines). Retail outlet: wine, cheese, breadsticks. Wine and Art Festival (June), Harvest Festival (October). Picnic tables in shaded grassy area near winery. Access for the handicapped. Directions: From Dayton, Interstate 75 west to Tipp City. Route 571 west to West Milton; left on Wheelock Road to winery.*

WISCONSIN

Cedarburg

Stone Mill Winery
N70 W6340 Bridge Road
Cedarburg, Wisconsin 53012
Telephone: (414) 377-8020
Owner: James Pape

Winery located in the Cedarburg Woolen Mills complex of shops in the historic Cedar Creek Settlement. The stone mill and surrounding village structures and sites are listed on the National Register of Historic Places, and include a covered bridge, lime kilns, the Ozaukee County Pioneer Village, Cedarburg Woolen Mills, and the Ozaukee Art Center. Stone Mill Winery produces 18,000 gallons of wine annually from purchased grapes and fruit. Product list includes the varietals Vidal Blanc, Niagara, and Catawba; and cherry, cranberry-apple, and strawberry fruit wines. Most noted for fruit wines. Guided tours of the winery (45 minutes) cover all aspects of wine production, and conclude with free tasting. Wine museum on premises. Open all year: Monday-Saturday, 10-5; Sunday, 12-5. Guided tours. Free tasting. Retail outlet: wine, gifts. Wine and Harvest Festival (third weekend in September), Grape Stomping Winter Fest (first weekend in February). Picnic facilities. (Restaurants nearby.) Access for the handicapped limited to wine shop and tasting room. Directions: From Milwaukee, north on Interstate 43 to Cedarburg exit. West on Highway C (Pioneer Road) 3 miles to Highway 57. North on Highway 57 to downtown Cedarburg. Stone Mill Winery in historic Cedarburg Woolen Mills on your right.

LaCrosse

Christina Wine Cellars
109 Vine Street
LaCrosse, Wisconsin 54601
Telephone: (608) 785-2210
Owner: Robert Lawlor

Attractive winery founded in 1979 by the Robert Lawlor family and named for oenologist daughter, Christina. Located in downtown LaCrosse near Riverside Park (at the confluence of the Mississippi and Black rivers) in an old railroad freight house, a building listed on the National Register of Historic Places. Boxcars parked on sidings provide additional retail space for the winery's extensive line of gifts. Tasting room richly furnished with antiques and noted for turn-of-the-century bar made for the Rock Island Brewery. No vineyards. Wines produced from purchased juice. Product line: Chablis, Burgundy, Octoberfest Wine, Catawba, Cranberry-Apple, Montmorency Cherry, Concord Grape, Niagara, Natural Apple, and Port. 8,000 gallon annual output. Most noted for fruit wines. Open Memorial Day-Labor Day: Monday-Sunday, 10-9. Guided tours daily at 1, 2, and 3 cover railroad depot, 1904 railroad car, winery; tour includes video. Free tasting. Retail outlet: wine, gifts. May Wine Fest (last weekend in May). Picnic facilities. Freight House Restaurant. Access for the handicapped. Directions: From downtown

LaCrosse, follow 3rd Street north towards Highway 53 and 35. At Vine Street, go west (left) towards the river (winery on your right).

Prairie du Sac

Wollersheim Winery
Highway 188 (P.O. Box 87)
Prairie du Sac, Wisconsin 53578
Owners: JoAnn and Robert Wollersheim

Wollersheim, Wisconsin's most historic and important winery, sits on a scenic ridge overlooking the Wisconsin River and Prairie du Sac and Sauk City. The winery was founded more than 125 years ago by the legendary Count Agoston Harazathy, who is often referred to as the father of American viticulture. Harazathy, who settled here in 1840 and founded what was to become Sauk City, planted a vineyard to his native European grape varieties on what is now the site of Wollersheim Winery. No one knows precisely why he left Wisconsin a few years later, but the state's winters and wet summers were not conducive to growing grapes and his vines died. Harazathy moved on to California's more hospitable grapegrowing climate, where he became San Diego's first sheriff and curator of the San Francisco Mint, but he is best remembered today as the founder of California's premium wine industry.

Wollersheim's handsome stone winery was built in 1858 by Peter Kehl, who modeled it after his family's winery in the Rhine Valley. After Harazathy's departure, Kehl took it over and replanted its vineyards. He and his family operated the winery successfully until 1899, when a particularly severe winter killed many of the vines and Kehl died. For nearly seventy-five years afterwards, the land and its buildings were used for more conventional farming purposes until they were purchased by JoAnn and Robert Wollersheim in 1972.

Today Wollersheim Winery is family owned and operated. (You're likely to see Robert Wollersheim in coveralls inspecting the vineyards when you visit.) The winery is one of the very few in Wisconsin that grows its own grapes, a success made possible by the mild climate effected by the moderating temperatures of the river, and the surrounding hills that protect the location from north winds. Around the historic cellars and winery, now listed on the National Register of Historic Places, the Wollersheims have planted twenty-three acres to French hybrids and two acres to viniferas. Grape varieties grown include Maréchal Foch, Léon Millot, Pinot Noir, Seyval Blanc, Ravat, and Johannisberg Riesling. From these grapes

Wollersheim

and additional purchased crops, the winery produces 10,000 gallons of wine each year (ninety-five percent still table wine, five percent sparkling wine). Wollersheim's labels include the varietal wines Seyval Blanc, Chardonnay, Johannisberg Riesling, and Maréchal Foch (all vintage dated), and the proprietary wines Sugarloaf (white), Domaine Reserve (vintage-dated estate-bottled red), and Ruby Nouveau. The winery is best known for its Domaine Reserve, Sugarloaf White, Ruby Nouveau, and Johannisberg Riesling.

Tours of the winery cover its rich past and architectural heritage and include a slide show on viticulture and winemaking. The historic house of the founding family and an old cave in the hillside—the site of the original wine cellar—are additional highlights. The cool underground limestone cellars house wine aging in varying sizes of old oak cooperage, some dating as far back as the Civil War. A tasting of Wollersheim wines, with commentary provided by a guide on the fine points of wine appreciation, ends the tour. If you choose to picnic on the patios surrounding the winery, you may, if lucky, see American bald eagles soaring over the property during your visit. These majestic birds nest and feed near Prairie du Sac's hydroelectric dam. To commemorate their presence, the winery recently introduced two new wines—Eagle Red and Eagle White. (One dollar from the sale of each bottle is contributed to the preservation of the birds' endangered roosting areas.)

Open all year, except New Year's Day, Easter, Thanksgiving, and Christmas: daily, 10-5. Guided tours ($2 per person). Free tasting with or without tour. Retail outlet: wine, winemaking supplies, gifts, cheese. Harvest Festival (second Sunday in October). Picnic facilities. Access for the handicapped. Directions: From Prairie du Sac, take Route 60 East, across the river, to Highway 188. Follow 188 South to winery (on your left).

Sturgeon Bay

Door Peninsula Winery
5806 Highway 42 North
Sturgeon Bay, Wisconsin 54235
Telephone: (414) 743-7431
Owners: Mark Feld and Tom Alberts

Fruit winery in Door County, one of the Midwest's most popular recreation areas. Located in Carlsville Schoolhouse (1868). Renovated for its current use in 1974. Features eleven wines made from fruit purchased from local orchards, including cherry, apple, plum, pear, cranberry, and strawberry wines among them. Guided tours (about twenty minutes) cover history of schoolhouse and winery, the various production areas, and short slide presentation. Tours end in the salon for a tasting of the full line of Door Peninsula wines and the winery's cherry wine cheese. Known for both dry and sweet cherry wines. Open all year. May-December: daily, 9-5; July-August: daily, 9-7. Guided tours (50¢ charge for adults). Tasting included in charge. Retail outlet: wine, gifts, snacks. Wine Festival (first Sunday in August). Picnic facilities. Limited access for the handicapped. Directions: From Sturgeon Bay, 8 miles north on Highway 42.

The Southwest

The Southwestern states of Texas and New Mexico represent an emerging new wine region that holds great promise for the future. Reflecting its natives' penchant for doing things in a big way, Texas is in the forefront of winegrowing in the area with fifteen wineries, many with all-vinifera vineyards. Beginning as early as the late 1800s, Texas had a wealth of wineries and vineyards, and was the home of one of America's most famous viticulturists and grape hybridizers, Thomas Munson. By 1974, though, only one Texan pre-Prohibition winery remained: the century-old Val Verde, tucked away in a southern corner of the state on the Rio Grande.

But in 1974, when a Texas A & M University study showed that most parts of the state were suitable for grapegrowing, there began a virtual explosion of vineyard planting. Texas A & M, for example, planted 2,000 acres to vineyards in the western part of the state. A group of investors has entered into agreement with the university to open a winery, Ste. Genevieve's, to vinify its crop.

There are now six recognized grapegrowing regions in Texas: South, North/Central, East, Hill Country, West, and the High Plains. All of the wineries and vineyards in these regions are roughly located in a wide horizontal swath cutting across the middle of the state. The best region for wine touring is the Hill Country, just north of Austin. Six wineries are located here, the most handsome of which is the new Fall Creek Vineyards in Tow. Ken Moyer's champagne cellar near the historic town of New Braunfels and Oberhellmann, in Fredericksburg are also well worth visiting.

Like the state's vineyards, Texan wine hospitality is still maturing. While many of its wineries offer tours and tastings, a "dry" mentality left over from Prohibition has put some restrictive wrinkles in recent state farm winery legislation and consequently in many a winery's ambience. In some counties where grapegrowing and winemaking are sanctioned, laws still prevent the purchase of wine or even the sampling of it on winery premises.

135

This rush to plant vineyards has also swept into neighboring New Mexico, where cheap land has attracted foreign as well as American investors. New Mexico's vinous history dates back to the 17th century with the making of sacramental wine by Spanish colonists. By the late 1800s, the state was the fifth largest wine producer in the country. Currently New Mexico has eight wineries and four wine regions: the "Four Corners" area where Utah, Colorado, New Mexico, and Arizona meet; the Mesilla Valley and the Las Cruces area, both in the south; and the area around Roswell, in the east. Of these wineries, there are two handsome ones near Albuquerque to visit: Anderson Valley, located on an Arabian horse farm on the city's outskirts; and West Winds, a twenty-minute drive north. Both are quite attractive, make vinifera varietal wines, and are very hospitable and instructive to tour.

NEW MEXICO

Albuquerque

Anderson Valley Vineyards
4920 Rio Grande Boulevard, N.W.
Albuquerque, New Mexico 87107
Telephone: (505) 344-7266
Owner: Patty Anderson

Family-owned-and-operated Anderson Valley Vineyards may be the only winery in the United States that shares acreage with an Arabian horse farm. But thoroughbred horses and premium wines are not the limits of the Anderson family's rather special pursuits. The adobe winery and its vinifera vineyard are owned by Patty Anderson, the widow of the late Maxie Anderson, who made headlines when he crossed the Atlantic Ocean in a hot-air balloon. (This achievement is reflected in the design of the winery's label and logo.) Winemaker and manager is Mrs. Anderson's son, Kristian.

The new winery (bonded in August, 1984) is situated in a pastoral setting along the Rio Grande River just outside Albuquerque. Its country setting, however, is anything but remote. Just hop on the trolley bus that stops at all the hotel entrances. The trolley is designed to take visitors to the local tourist attractions, and Anderson Valley Vineyards is a stop on its itinerary. After the tour and tasting, flag the trolley again and you can ride it to one of the area restaurants for dinner.

Anderson Valley's fifteen-acre vineyard is planted to Chenin Blanc, Chardonnay, and Johannisberg Riesling. From these grapes and additional fruit purchased from New Mexico and California growers, it produces varietal wines and a generic burgundy. Anderson Valley takes special pride in its Johannisberg Riesling and Chardonnay.

Guided tours of the attractive winery are given throughout the year; they last approximately twenty minutes and are tailored to the interests of the participants. You can taste Anderson Valley's whole line in its handsomely tiled tasting room. The ballooning interests of the Anderson family are reflected in many of the gift items available. In addition to the latter, there is a good selection of local crafts for sale, including handcrafted pottery and baskets.

During the day you can picnic on the winery's patio or in the nature preserve along the river, a short walk away.

Open all year, except major holidays: Tuesday-Saturday, 12-5:30. Guided tours. Free tasting. Retail outlet: wine, local crafts. Picnic facilities. Access for the handicapped. Directions: From Albuquerque, take Interstate 40 to Rio Grande exit. Follow Rio Grande north for 3 miles to winery (on east side of road). Or take the trolley bus, if you prefer.

Bernalillo

Westwind Winery
N.M. Highway 44 (P.O. Box 786)
Bernalillo, New Mexico 87004
Telephone: (505) 867-3000
Owner: James R. Winchell

Small premium winery (6,000 cases annually) fifteen minutes north of Albuquerque. New 10,000-foot award-winning winery building designed by builder/owner, James Winchell. UC Davis-trained winemaker produces award-winning wines. All grapes purchased from New Mexico growers. Experimental vinifera and French-American hybrid vineyard on winery site.

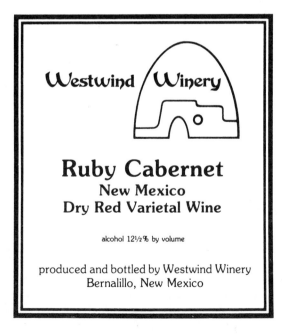

Product list primarily dry table wines, including vintage-dated varietals (some are vineyard-designated, as well): French Colombard, Vidal Blanc, Ruby Cabernet, and Zinfandel. Proprietary blend: Rio Grande Rojo. 1983 Ruby Cabernet won blue ribbon for best commerical red wine in New Mexico. Open all year: Monday-Saturday, 10-4. Guided tours. Free tasting. Retail outlet: wine, gifts. Harvest festival (October). Picnic facilities planned. Directions: From Albuquerque, Interstate 25 north to N.M. Highway 44. Highway 44 west to winery.

Las Cruces

Binns Vineyard and Winery
3910 West Picacho Avenue
Las Cruces, New Mexico 88001
Telephone: (505) 522-2211
Owner: Eddie Binns

Small estate winery in Las Cruces. Owned by Eddie Binns; his son, Glen, is winemaster. Established in 1982. Twenty-acre vineyard planted to Chenin Blanc, Zinfandel, French Colombard, Johannisberg Riesling, and Chardonnay. Yield supplemented by grapes purchased from Mesilla Valley growers. 10,000 cases annually. Best known for Riesling and blush Zinfandel. Guided tours of winery by appointment only. Free tasting and retail sales in new tasting room. Nice park for picnicking on winery premises; many good restaurants nearby. Open all year: Monday, Wednesday, and Friday, 1-6. Guided tours by appointment only (two days' notice suggested). Free tasting. Retail outlet: wine, gifts. Fiddling Contest (June); Whole Enchilada Fiesta (October). Picnic facilities. Access for the handicapped. Directions: From Las Cruces, west on Highway 70 (also called West Picacho Avenue) to winery.

TEXAS

Bryan

Messina Hof Wine Cellars
Old Reliance Road (Route 7, Box 905)
Bryan, Texas 77802
Telephone: (409) 779-2411
Owners: Paul and Merrill Bonarrigo

Paul Bonarrigo founded Messina Hof Wine Cellars as the realization of a lifelong dream. His family has been making wine for its own consumption for more than a century, and for all of that time the first-born son of the new generation (always named Paul) has been trained as the winemaker. Bonarrigo named his new winery, established in 1983, in honor of that tradition (his ancestors emigrated to the United States from Messina, Sicily, in the 1920s) and for his winemaker, Ron Perry, whose family came from Hof, Bavaria.

Located three miles from Bryan, Messina Hof is eastern Texas's only vineyard/winery. Planted on its thirty-acre vineyard are the grape varieties Cabernet Sauvignon, Chenin Blanc, Muscat Canelli, Cabernet Franc, Johannisberg Riesling, and Lenoir. From these grapes, Messina Hof produces 18,000 gallons of wine annually, eighty percent of which is table wine, twenty percent fortified wine. The product list features such vintage-dated varietal table wines as Cabernet Sauvignon, Johannisberg Riesling, and Chenin Blanc; the generic table wines Vino Di Amore, Sweet Bianco, and Rosso; and fortified wines, among them Papa Paulo Porto. It is particularly known for the latter and for its white Zinfandel and Vino Di Amore.

Guided tours of the winery are given by appointment on the third Saturday of each month, and more frequently when Texas A & M's football team is playing at home. Tours are conducted by Paul Bonarrigo and cover both vineyard and production facility. A slide presentation on winemaking is included. Following the tour you can usually sample four of Messina Hof's wines in the tasting room which adjoins the barrel-aging cellar. There are picnic and barbeque facilities next to the winery. And if luck is with you you might catch a bass or catfish in the nearby lake to grill for lunch.

The current winery building is somewhat primitive, but plans call for a new reception and tasting room to be installed in the French manor

140

house which is being reassembled on the property. It will combine the winery's hospitality rooms with bed and breakfast accommodations.

Open all year, except first week in January: Monday-Friday, 8-5. Guided tours by appointment (one day's notice required): third Saturday of each month and weekends of Texas A & M home football games, 11-5. Free tasting. Retail outlet: wine, gifts. Wild Flower Festival (Spring). Picnic and barbeque facilities near lake. Fishing. Access for the handicapped. Directions: From Houston, Highway 6 to Booneville Road exit. Access Road north to Old Reliance Road. Old Reliance Road east 2½ miles to winery.

Del Rio

Val Verde Winery
139 Hudson Drive
Del Rio, Texas 78840
Telephone: (512) 775-9714
Owner: Thomas M. Qualia

The border town of Del Rio, located in the southernmost part of Texas, boasts the state's oldest winery, Val Verde. Founded in 1883 by Italian immigrant Frank Qualia, it is owned and operated today by his grandson Thomas, known locally as Tommy. The winery is located in the original thick-walled, 19th-century Spanish-style adobe structure that the elder Qualia built. The building, an officially designated historic site, is shaded by pecan and oak trees, and inside you will see much of the original winemaking equipment, including the original press Frank Qualia shipped from Italy.

In the tradition that Tommy's grandfather started and his father, Louis, continued, Val Verde is still a family operation run largely by Tommy as viticulturist, winemaker, and even part-time plumber, when needed. Tommy's wife, Linda Kay, and oldest son, Michael, lend a helping hand, too, especially with the manual bottling line. Val Verde is thus operated on the small-estate principle, with almost all of the grapes used in production grown in its own vineyards. While respecting his family's winemaking history and tradition, Tommy Qualia has kept pace with the latest winemaking practices. These are reflected in the winery's newly added laboratory facilities, modern fermentation tanks, and experimental vinifera vineyard.

Val Verde currently has thirty acres under cultivation planted to Lenoir and Herbemont, native American varieties. In addition, an eight-acre experimental plot is planted to vinifera. Twelve of the vineyard's acres are located behind the winery; the remainder, thirty miles south of town. Val Verde produces 6,000 gallons of wine annually, in six varieties: Herbemont (dry), Herbemont (semi-sweet), Rosé of Lenoir, Lenoir, Tawny Port, and Johannisberg Riesling. The last was introduced in honor of Val Verde's centennial in 1983 and is made from grapes purchased in Peco County. Rosé of Lenoir is a silver medal winner. Other award winners are the Lenoir and Tawny Port.

Tours of the small winery are given on request (ask the tasting room staff). Val Verde wine is sold here, along with wine accessories, gifts, books, and fresh juice and grapes in season (August).

Open all year: Monday-Saturday, 9-5. Guided tours on request. Free tasting. Retail outlet: wine, wine-related items. Access for the handicapped. Directions: From Highway 90 in Del Rio, cross overpass in town. Go right on East Gibbs to Pecan. At Pecan, left to winery. (There are signs along the way.)

Fredericksburg

Oberhellmann Vineyards
Highway 16 (Llano Route, Box 22)
Fredericksburg, Texas 78624
Telephone: (512) 685-3297
Owner: Oberhellmann, Inc.

Oberhellmann Vineyards was founded in 1974 by Austin food broker Robert Oberhelman. (Oberhelman spells his own name with one "l" and one "n", but to emphasize his Germanic roots, he spells the winery's name with two of each.) The winery is one of six winegrowing estates in the Hill Country, a region which extends for about eighty miles northwest from San Antonio and Austin. Located fourteen miles north of Fredericksburg on Highway 16 (known locally as Llano Road), the winery's old-world design reflects its owner's German heritage. It is surrounded by a thirty-acre all-vinifera vineyard which is planted to Chardonnay, Johannisberg Riesling, Gewürztraminer, Sauvignon Blanc, Sémillon, Pinot Noir, and Cabernet Sauvignon. From these grapes, winemaker Oberhelman produces 12,000 gallons of varietal wines each year. His emphasis is primarily on white wines, such as Chardonnay, Gewürztraminer, and Johannisberg Riesling, but he is also establishing a reputation for Cabernet Sauvignon and Pinot Noir.

Short (twenty-minute) guided tours of the winery are given on Satur-

days (the only day the winery is open to the public). Afterwards, over a glass of Oberhellmann wine in the tasting room, you can study the photos displayed on its walls which further illuminate the process which transforms grapes into wine.

Open May to mid-December: Saturday only, 10-5. Guided tours at 11, 1, and 3. Free tasting. Retail outlet: wine, wine accessories, and books. Directions: From Fredericksburg, north on Highway 16 for 14 miles to winery (on your right).

Lubbock

Llano Estacado Winery
Farm Road 1585 (P.O. Box 3487)
Lubbock, Texas 79452
Telephone: (806) 745-2258
President: C. M. McPherson

In the 1540s Francisco Vasquez de Coronado explored this area of the Southwest in search of the fabled "seven cities of gold," marking his route through the tall buffalo grass with wooden stakes so that he could find his way back. Hence this region of Texas, and the winery established here, came to be called Llano Estacado, which means "staked plains" in Spanish. Llano Estacado was the first winery bonded in the Lone Star State in this century. Its large white production facility and hospitality center, located three miles south of Lubbock, opened in 1976. And the winery has since emerged as one of the leaders in Texas's premium wine industry.

Surrounding the winery is Llano Estacado's ten-acre all-vinifera vineyard, planted to Chenin Blanc and Johannisberg Riesling. The owners, Texas Tech faculty members Clinton McPherson, Roy Mitchell, and Robert Reed, supplement this crop with vinifera grapes provided by area growers. (Purchased grapes account for eighty percent of the grapes needed for production.) McPherson's son, Kim, is the winemaker. He was educated at UC Davis and afterwards apprenticed at Trefethen Vineyards in California.

Llano Estacado annually produces 25,000 cases of both vintage-dated varietal wines (the Llano Estacado label) and generics (the Staked Plains label). The emphasis is on the former, which have won many awards. Varietal whites on Llano Estacado's product list are Chenin Blanc, Johannisberg Riesling, Chardonnay, Gewürztraminer, French Colombard, and Sauvignon Blanc. Among the red varietals produced are Zinfandel and Cabernet Sauvignon; table wines are Mesa Rouge and Mesa Blanc.

Guided tours of the winery are given on weekends. The modern facility is designed so that you can view much of the winery's operation from its tasting room. But the half-hour tour will give you a closer look at the vineyard and at production methods, including the fermentation room and state-of-the-art bottling line.

Open all year: Saturday, 12-6; Sunday, 1-6. Guided tours. Free tasting. Access for the handicapped. Directions: From Lubbock, Highway 87 south to Farm Road 1585. East on 1585, 3½ miles to winery.

Pheasant Ridge Winery
1505 Elkhart Avenue (Route 3, Box 191)
Lubbock, Texas 79416
Telephone: (806) 746-6033
Owners: C. Robert Cox Jr. and Charles Robert Cox III

PHEASANT RIDGE

TEXAS
CABERNET SAUVIGNON
1982

LUBBOCK COUNTY
Produced and Bottled by Pheasant Ridge Winery, Lubbock, Texas   Alcohol 11.8% by Volume

Small estate winery 3 miles southeast of New Deal in Lubbock County. Family owned and operated. Thirty-four-acre all-vinifera vineyard planted in 1979 to Chenin Blanc, Sauvignon Blanc, Cabernet Sauvignon, Merlot, Cabernet Franc, Ruby Cabernet, and French Colombard. Winery built in 1982; bonded the same year. Production: 2,200 cases of wine annually; only varietal table wines. Best noted for 1982 Cabernet Sauvignon which is aged in Nevers and French-coopered American oak. Winery is named for the ring-necked pheasants which roost in the shelter of the trellises. Open all year by appointment only. Guided tours. Free tasting. Picnicking under the grape arbor. Directions: From Lubbock, north on Highway 87 to Farm Road 1729. Follow Farm Road 1729 east for 2 miles to Elkhart Avenue. At Elkhart south for one mile to winery.

New Braunfels

Moyer Texas Champagne Company
1941 Interstate Highway 35 East
New Braunfels, Texas 78130
Telephone: (512) 625-5181
Owners: Ken and Mary Moyer

Located near the German-settled town of New Braunfels in the Hill Country wine region. Founded in 1980 by Ken and Mary Moyer who also own the Moyer Vineyards Winery and Restaurant in Manchester, Ohio. Winery produces one of the state's first methode champenoise *sparkling wines. All grapes purchased from California vineyards. Output: 10,000 gallons annually. Open all year: Monday-Saturday, 10-5. Guided tours. Free tasting. Retail outlet: champagne. Access for the handicapped. Directions: On Interstate 35 East, south of Austin.*

Springtown

La Buena Vida Vineyards
WSR Box 18-3
Springtown, Texas 76082
Telephone: (817) 237-9463
Owner: Dr. Bobby Smith

Established in 1974, La Buena Vida Vineyards is one of three winery/

vineyards located in Parker County. All can be considered anomalies, since the county's residents have consistently voted to maintain its "dry" status since it was first instituted in 1913. That the wineries exist here at all is due to the passage of Texas's farm winery act, which Buena Vida owner Bobby Smith was instrumental in helping pass. This state-wide legislation permits winegrowers to process wine in dry counties, though each local voting precinct maintains the right to decide whether to permit its sale. Because Parker County's residents continue to refuse vintners that privilege, Bobby Smith and his colleagues can neither sell nor serve their wines at their vineyards. La Buena Vida, however, maintains a tasting room twenty miles away in Lakeside (over the border in Tarrant County). Its vineyards are open by appointment; you might like to tour them first and then visit the tasting room to sample the product and to see a narrated slide presentation on the winemaking process.

Family-owned-and-operated La Buena Vida is one of the state's leading premium wineries. Planted in its twelve-acre estate vineyard are French-American hybrids and viniferas from which 20,000 gallons of wine are produced annually. The winemaker is UC-Davis-trained Steven L. Smith, who is aided by his father, Bobby, also a UC-Davis alumnus.

Visitors to La Buena Vida's tasting room in Lakeside may sample all of the winery's fourteen vintages. Buena Vida is particularly known for its Rayon d'Or (a gold-medal winner), Texas vintage port, and Blanc de Noirs, a *methode champenoise* sparkling wine.

Winery open by appointment only. Tasting room (Lakeside) open all year: Monday-Saturday, 10-5; Sunday, 12-5. Narrated slide presentation on winemaking. Free tasting. Retail outlet: wine, wine accessories and wine-related gifts. Texas Wine Country Chili Cook-off (May). All Texas Grape Stompin' Contest (September). Access for the handicapped. Directions to tasting room: located 10 miles from downtown Fort Worth on Highway 199, 2 miles northwest of the Lake Worth Bridge at the FM-1886 cutoff. Call for directions to winery. (Mailing address: Route 2, Box 927, Fort Worth, Texas 76135.)

Tow

Fall Creek Vineyards
Highway 2241
Tow, Texas 78672
Telephone: (512) 476-4477
Owners: Hugo and Susan Auler

If you're visiting the Hill Country wineries, you won't want to miss Fall Creek Vineyards. For it is the largest of the winegrowing estates in this prolific vinous area of Texas; has an attractive facility designed by co-owner Susan Auler, who holds a degree in interior design; and occupies an especially scenic location on the shores of Lake Buchanan. Its wines have won numerous awards at the Texas State Fair each year since 1982.

The combined winery and ranch of the Auler family is a handsome, large two-story structure whose exterior design reflects the early architecture of the Austin area: a graceful archway embellishes its facade and a brick entrance courtyard faces the vineyards. Forty-five acres are planted to Sauvignon Blanc, Chardonnay, Chenin Blanc, Emerald Riesling, Cabernet Sauvignon, Carnelian, and Zinfandel. The winery and its vineyard occupy land that has been in the Auler family for four generations; across the road is their Fall Creek Ranch, where you might spot a herd of cattle peacefully grazing.

Fall Creek Vineyards, with a 33,000-gallon annual output, produces six wines, including the vintage-dated varietals Emerald Riesling, Chenin Blanc, Sauvignon Blanc, and Carnelian. Its proprietary blends include Proprietor's Red and Blanc de Blancs.

Guided tours of the modern winery are available only on the last Saturday of each month, unfortunately the only time Fall Creek is open to the public. While there are no picnic facilities on the premises, you can take your hamper to the lovely public park located nearby on the shores of Lake Buchanan.

Open January-October: last Saturday of each month, 1-5. Guided tours. Free tasting. Retail outlet: wine accessories (no wine sold, as winery is in a dry precinct.) Access for the handicapped. Directions: In Llano County between Llano and Burnet. 2.2 miles northeast of Tow post office on Highway 2241. (Mailing address: 1111 Guadalupe, Austin, Texas 78701.)

California

The golden state of California far and away leads the nation in grapegrowing and winemaking. At last count it had 739,000 acres planted to vineyards—more than five times that of the other wine-producing states combined—harvested five million tons of grapes annually, and produced seven-tenths of the wine consumed in America. Its bounteous vineyards also provide vintners in other states and Canada with grapes for wine.

California's wineries also outnumber all other states and provinces. At last count there were more than 600. According to California resident Leon Adams, author of *The Wines of America,* "it is easier to raise vinifera grapes in the fabulous climate of California than anywhere else in the world." And, "its climate," he continues, "is the envy of European winemakers." California, unlike France, rarely has a bad year. And, consequently, vintage dating on its labels is more for the identification of different lots of wine rather than indicative of a particular growing season. Wine production in the state is also not confined to the well-known Napa and Sonoma Valleys, as an out-of-stater might be led to believe. California's 12,000 vineyards, covering 1,000 square miles, are located in forty-one of its forty-eight counties from as far north as Lake County to as far south as San Diego.

Wine touring in California is almost limitless. There are old wineries with century-old stone caves tunneled into the earth (such as Schramsberg) and slick, new high-tech wineries like Sonoma Cutrer with the only "breathing" earth cellar in North America. There are large and small family wineries, including Sebastiani, with its unparalleled collection of carved casks, and charming Glen Ellen, run by the extended Benziger family. There are monumental church-owned wineries—the Greystone Cellars being among the most famous—fashionable estate wineries such as Iron Horse, and even corporate-sized American and European holdings such as Seagram's (Sterling Vineyards) and Moët Hennessey (Domaine Chandon).

148

In California you will discover almost every type of wine produced in North America, as well as an extraordinarily rich selection of winery architecture. The wineries vie with each other to provide the best in visitor amenities and educational facilities. Where else can you ride a tram to a hilltop winery (Sterling), dine in a winery's three-star restaurant (Domaine Chandon), take a cooking class with Julia Child (Robert Mondavi), savor a barbeque lunch grilled over vine cuttings (Callaway), or enjoy Shakespeare amid the vines (Buena Vista)?

The following selections represent only a small, but noteworthy, sampling of the possibilities afforded the California visitor; to cover all the wineries that welcome visitors would fill a whole book. For ease in planning trips, the wineries have been divided into five geographic regions: Napa Valley, Sonoma Valley and County, Mendocino and Lake Counties, Bay Area and North Central Coast, and South Central Coast and South Coast. As you will discover, the greatest concentration of wineries, and some of the most spectacular, are located in the Napa Valley along Highway 29 and the parallel Silverado Trail, from Napa north to Calistoga. Unlike their counterparts in the rest of the United States and Canada, many of these wineries can be visited by appointment only, since the smaller ones would be swamped by large groups of tourists unless such restrictions were imposed.

Neighboring Sonoma Valley's wineries, closer to San Francisco, are more spread out, as are the wineries in the rest of the state, including those of Lake and Mendocino Counties, and the Bay Area and North Central Coast. Further south are newer wine regions where grazing lands have provided fertile ground for vineyards and wineries that have already gained recognition.

NAPA VALLEY

Calistoga

Chateau Montelena Winery
1429 Tubbs Lane
Calistoga, California 94515
Telephone: (707) 942-5105
Owners: James L. Barrett, Leland Paschich, Ernest Hahn

Chateau Montelena, the most northerly of the Napa Valley's many wineries, is somewhat off the beaten track, but worth the trip. Not only is the winery one of the valley's most picturesque, but you can "take the cure" along the way at one of the numerous spas in the quaint town of Calistoga, famous for its mud baths, mineral waters, and geyser. (The latter faithfully erupts on the half hour.)

Located on Tubbs Lane a few miles east of the town, the winery is housed in a handsome stone castle built into the foot of a hill by its founder Alfred L. Tubbs around 1882. Tubbs, a prominent San Francisco figure at the turn of the century, built the chateau in the French style. Its front is faced in imported cut stone and the walls range from three to twelve feet in thickness and extend into the hillside, keeping the cellar naturally cool and damp even on the hottest summer days. In keeping with the architecture, Tubbs planted only select varietal vine cuttings which he imported from France along with a French winemaker, Jerome Bardot. The wines this team produced enjoyed considerable renown during this early period of California winemaking.

During Prohibition and for some years after, the vineyards and winery fell into disuse and neglect, but in 1972, under the new ownership of James L. Barrett, Lee J. Paschich, and Ernest W. Hahn, the property began to "hum" again—literally. The present owners are great believers that without a spirit of caring no great wine can be made. They claim to sing and hum to the wine and that it responds and becomes happy. Aside from this mystical ingredient, they have also implemented the latest state-of-the-art technology such as membrane filters, along with an extensive barrel-aging program which includes French Limousin and Nevers oak.

Today, the deceptive approach to Chateau Montelena hides its lovely facade. From the parking area the path leads to the tasting and salesroom entrance which is located on the side of the chateau. Here the tour begins. (These are by appointment only, to keep the groups as small as possible.) Tours generally focus on the season and what is happening in the vineyard

and winery at that time of year. As the group winds down the hill to the front of the chateau, the winery's interesting history and architecture are discussed, and another good reason to visit Chateau Montelena is seen: At the foot of the chateau are formal Chinese gardens complete with a lake, pagoda, island, and a beached junk, all added by former Chinese owners. Beyond, Mount St. Helena looms majestically with meticulously tended vineyards in the foreground. The tour continues around the lake, going into the vineyard, through the winemaking facility and aging cellar, and ending up in the tasting room.

Usually two of the Chateau's four wines are available for tasting daily—a red and a white. (The other two are opened on alternate days.) Its product line of 100 percent varietal table wines includes a Johannisberg Riesling, Chardonnay, Zinfandel, and Cabernet Sauvignon. Chateau Montelena is best known for its Chardonnay. Its 1973 vintage put the winery on the map. It was awarded first place in the Paris Tasting of May 1976 over nine other top-of-the-line French white Burgundies and California Chardonnays in a blind tasting by a panel composed exclusively of prestigious French wine notables. Chateau Montelena is also very proud of its Cabernet, since the grapes for this fine wine are grown in the surrounding vineyards.

Open all year: daily 10 - 4. Guided tours (appointments necessary) at 11:00 and 2:00, approximately 45 minutes in duration. (Call or write as far ahead as possible.) Free tasting. Retail outlet: wine and gifts. Picnicking on grounds by strict appointment only. Access for the handicapped. Directions: 128 north to Tubbs Lane. Right on Tubbs Lane past geyser to winery on left.

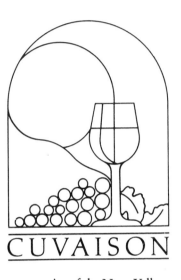

Cuvaison Vineyard
4550 Silverado Trail
Calistoga, California 94558
Telephone: (707) 942-6266
President: Robert Logan

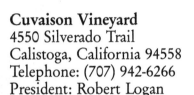

Cuvaison, just south of Calistoga on the eastern rim of the Napa Valley, is a small winery by California standards—20,000 cases annually. It is also relatively new. It was founded in 1969 by a group of engineer/scientists. A few years later it was purchased by a New York publishing company that built the present facility, and it is now owned by a Swiss firm that installed Dr. Robert Logan, a former professor of viticulture and oenology at UC Davis, as president.

The winery is located in a lovely Mission-style building complete with red-tiled roof, white walls, and curved arches shaded by massive oak trees. Cuvaison makes three varietal table wines: Chardonnay, Cabernet Sauvignon, and Zinfandel, but is particularly noted for its Chardonnay which is fermented and aged in small French oak barrels with a portion left in stainless-steel cooperage.

The tours, by appointment only, are fifteen to thirty minutes in length and feature the winery's beautiful stained-glass windows, which were created by a local artist. These are also reflected in Cuvaison's distinctive wine label. Tastings of all three wines are available daily.

Open all year: daily 10 - 4. Guided tours, by appointment, Saturday and Sunday. Free tasting. Retail outlet: wine and wine-related items. Picnic facilities. Access for the handicapped. Directions: From St. Helena, Highway 128 north to Dunaweal Lane. Right on Dunaweal to Silverado Trail. Right turn on Silverado Trail. Cuvaison on left. Mailing address: P.O. Box 2230, Napa, California 94558.

Hanns Kornell Champagne Cellars
1091 Larkmead Lane
Calistoga, California 94515
Telephone: (707) 963-1237
Owner: Hanns Kornell

If you're on the trail of sparkling wine, you may want to stop at Hanns Kornell Champagne Cellars. While the tasting room here is not as attractive as some (it's due to be renovated soon) and there are no vineyards, (Mr. Kornell produces all of his sparkling wine from *cuvées* he makes from "finish" wine purchased from neighboring wineries), the tour is very complete, the owner and his family hospitable, and a free tasting of their sparkling wine well worth the visit.

The tour, which is occasionally guided by a family member, begins at the tasting room, an unimposing white frame building, and continues across the courtyard to the handsome original stone cellar with wonderful thick walls and high ceilings—now a California historic landmark. This structure serves as an aging facility. Here riddling is discussed and demonstrated. All riddling is done by hand since Mr. Kornell feels one should handle the bottles as gently as possible. Visitors are then given an overview of the champagne-making process. Champagne is produced here in the traditional French *methode champenoise*, i.e., fermented in the bottle. The history of the winery is covered and a bit of Mr. Kornell's rags-to-riches past is touched upon. Next the tour moves to the bottling area where you can see both hand- and fully-automated (depending on the bottle size) disgorging, dosage, corking, labeling, etc. The tour, which ends up in the tasting room, takes about twenty minutes. Usually two champagnes can be tasted daily.

Kornell's product line includes four dry champagnes—Blanc de Blancs, Sehr Trocken, Brut, Extra Dry—and a sweeter red champagne, Rouge.

Open all year, daily except major holidays: 10 - 4:30. Guided tours at 10 and 4, approximately 20 minutes in duration. Free tasting. Retail store. Access for the handicapped. Directions: four miles north of St. Helena on Highway 29 to Larkmead Lane. Right turn on Larkmead. Mailing address: P.O. Box 249, St. Helena, California 94574.

Schramsberg Vineyards
Schramsberg Road
Calistoga, California 94515
Telephone: (707) 942-4588
Owners: Jack and Jamie Davies

Although this well-known winery offers no tastings (because of limited production), a visit to Schramsberg is definitely a must for champagne lovers. Not only is the setting idyllic and steeped in history, but the winery's product is famous as well. (Schramsberg's Blanc de Blancs was the champagne both former-President Nixon and President Reagan took with them as gifts on state visits to Peking.)

The winding private road up the hill to Schramsberg from Highway 29 suggests "the forest primeval." Ducks bathe in a tranquil pond to the left, and the sun filters through the dense vegetation and huge trees along the way. At the end of the lane in a clearing stands the Victorian house that the winery's founder, Jacob Schram, built in the late 1800s. (Jack and Jamie Davies, the current owners, now make it their home. And this is one of the reasons for the "by appointment only" visitor policy.) The house and its 200-acre estate is also a registered California Historical Landmark since Robert Louis Stevenson, who stayed here in 1880 as a house guest of Mr. Schram's, later immortalized the estate by describing it in a chapter of his novel *Silverado Squatters*. To the right is a newer structure, the winery and offices. And to the left are the famous cellars—what everyone comes to see—a 1½-mile network of tunnels chipped out of volcanic ash by Chinese laborers nearly 100 years ago.

The tour, which takes about forty minutes, covers the complete production facility for making sparkling wine in the *methode champenoise*—fermentation, riddling, dosage, bottling line, etc.—and ends up in a small retail area. But the caves which house nearly 1.5 million bottles of the sparkling wine are surely the highlight of the tour.

Schramsberg's vintages include five different styles of champagne: Blanc de Blancs, which is a combination of two white grapes, Chardonnay and Pinot Noir; Blanc de Noirs, made primarily from the grape Pinot Noir; Cuvée de Pinot, a dry rosé; Cremant Demi-sec, a dessert champagne made in the classic cremant style; and Reserve, a limited bottling, aged over four years and representing the choicest of the harvest and the best winemaking efforts of the vineyard.

Open all year, by appointment only. Guided tour, 40 minutes. Retail store. Access for the handicapped. Directions: Take Highway 29 approximately 1/5 mile north of Larkmead Lane (the entrance to Hanns Kornell). Left on Peterson Lane. Follow sign; road jogs right and then up the hill to Schramsberg.

Sterling Vineyards
1111 Dunaweal Lane (P.O. Box 365)
Calistoga, California 94515
Telephone: (707) 942-5151
Owner: The Seagram Classics Wine Company

On a clear day you can see forever . . . or so it seems from Sterling Vineyards' hilltop location. The views of the Napa Valley are spectacular from the winery terraces, where on warm days you can bask in the sun and watch the glider planes taking off from nearby Calistoga, two miles north. This unique winery is the most dramatic in the Napa Valley and one of its most popular attractions. (Approximately 200,000 guests visit Sterling each year.)

A delightful consequence of the unique location is a four-minute tram ride to the winery from the parking lot on the valley floor 300 feet below. Passing through oleanders, mountain fir, and rosemary bushes, you eventually climb to the handsome building which dominates the southeast slope with its stark white Mediterranean-style architecture, belfrey, and steep stairways. Once there, the tour is at your leisure, since it is self-guided and you follow public viewing galleries punctuated with excellent informative signs, graphics, and color photographs. One feature of Sterling Vineyards is immediately apparent: the main cellar takes advantage of

the winery's natural terrain in that it runs downhill from the crusher, press, and stainless-steel fermenting and storage tanks through a series of oak barrel aging cellars. The purpose of this tiered design, dictated in part by the slope of the hillside, is to permit the use of gravity flow rather than pumps to move the wine during its development. Another highlight of the tour is a two-story cellar, or chai, which holds nearly 1200 French oak barrels used for aging the firm's reserve wines.

The visit ends in the visitors' center, a space dominated by a light-filled atrium. In the adjacent Dunaweal Room, the main tasting area, Sterling's current releases can be sampled. Here the tasting room staff is available to answer any questions raised by the tour or about the wine.

Sterling owns over 500 acres of vineyards in the Napa Valley, planted with 200,000 vines currently yielding 2,100 tons of grapes. The firm produces only vintage-dated, estate-bottled varietal wines and is particularly noted for its Cabernet Sauvignon, Merlot, Chardonnay, and Sauvignon Blanc.

By the time you are ready for the return trip on the tram, you most likely will have heard the Sterling bells. These toll on the quarter hour and were brought to California from the Church of St. Dunstan-in-the-East, London, England, where the bells were installed in the early 1700s. (The church itself was destroyed in World War II and the bells removed and recast. They were brought to the winery in 1972.)

Open daily: April 1 - December 31, 10:30-4:30; closed Sunday and Monday: January 1 - March 31. Tours, self-guided, approximately 45 minutes in duration. Free tasting. Retail store: wine and wine-related items. Tram: $3.50. Picnic facilities. Van available to transport handicapped to top of hill. Tasting area accessible, but tour difficult. Directions: Seven miles north of St. Helena on Route 29 on right. (You can't miss it!)

Napa

Clos Du Val
5330 Silverado Trail
Napa, California 94558
Telephone: (707) 252-6711
Owner: John Goelet

Clos du Val, located on the Silverado Trail, looks like a classic French chateau with touches of California. The vineyards and winery are owned by John Goelet, the French-American millionaire-owner of Chemical Bank. He founded the winery in 1972 on the advice of Bernard Portet, now its winemaker and general manager. Mr. Goelet, who comes from a family of important Bordeaux wine merchants, had hired Mr. Portet to search out an area of the world capable of producing wine as good as that made in France, and the search ended in the Napa Valley.

A native of the Cognac region of France, Mr. Portet comes from six generations of winemakers. He spent most of his childhood in one of the finest wine-producing regions in the world. His father, André Portet, was the Regisseur (technical director) of Chateau Lafite-Rothschild for twenty-one years and the family moved to the estate when Bernard was seven.

Clos du Val (which means "small estate in the valley") is set against the vine-covered hills of Stags' Leap. In typically French fashion, roses mark each row of grapevines which are planted on a slight slope facing the setting sun. The vineyards, which surround the winery, are planted to Cabernet, Merlot, and Zinfandel. Pinot Noir and Chardonnay are planted in newer vineyards in the Carneros Creek region.

The winery is particularly noted for its Cabernet Sauvignon, Zinfandel, Merlot, and Pinot Noir and for Bernard Portet's personal style of winemaking, which reflects a shift away from the current trend of purely varietal wines. (For example, Mr. Portet tends to mix his Cabernet with Merlot to ensure a fine-tasting wine. "I could make a Cabernet that is more interesting," he adds, "but first of all I want a wine that tastes good. To me, taste is more important than the nose.")

Tours at Clos du Val are tailored to the sophistication of the participants. But they usually cover the philosophy, history, property, and facilities (including the aging cellar, vineyard, and production operation) followed by a commented wine tasting in the retail sales area. (The walls here are decorated with an unusual collection of original and amusing Ronald Searle wine-motif cartoons. Reproductions of these are available on posters, postcards and playing cards.)

There are limited picnicking facilities under the shade trees along the entrance road, but the staff can recommend other nice spots to picnickers they can't accommodate.

Open all year, daily except major holidays: 10 - 4. Guided tours: 10 and 2 daily during summer months, otherwise by appointment only (one day's notice needed). Free tastings. Retail outlet: wine, gifts. Picnic tables (shaded by a California oak). Access for the handicapped. Directions: Five miles north of Napa on east side of Silverado Trail.

Mayacamas Vineyards
1155 Lokoya Road
Napa, California 94558
Telephone: (707) 224-4030
Owners: Bob and Nonnie Travers

Spectacular mountain location, with views on clear days as far as San Francisco. Vineyards planted primarily to Cabernet Sauvignon. Primary wines: Chardonnay and Cabernet, both vintage dated. Guided tours, sales, tasting, by appointment only. Directions: Highway 29 north from Napa. Left on Redwood/Mt. Veeder Road. A steep drive up Mt. Veeder to Lokoya.

Monticello Cellars
4242 Big Ranch Road
Napa, California 94558
Telephone: (707) 253-2802
Owner: Jay Corley

Amid a 200-acre plantation of vineyards in the southern reaches of the Napa Valley is a surprisingly un-Californian piece of architecture that will stir the memory of every student of American history. It is the home of Monticello Cellars, and, while *this* Monticello is not a line-for-line copy of the famous Jeffersonian homestead, it bears an unmistakable resemblance to the original.

The resemblance of the new winery to the Virginia landmark is no coincidence since the winery's owner, Jay Corley, hails from that state. And it is his hope that a winery building and philosophy so conceived will remind wine connoisseurs that the third president of the United States was, among other things, a great innovator and contributor to American viticulture and cuisine. Before he decided to build his own winery, Mr. Corley was a grower of Chardonnay, one of the largest in the Napa Valley, and sold his grapes to some of the most respected premium wineries.

Visitors to Monticello Cellars tour the vineyards and the working winery with its de-juicing tanks, presses, centrifuge, and fermentors, and then walk across to the underground aging cellars burrowed under Monticello, where a handsome tasting room finished with 100-year-old redwood panels from dismantled wine casks awaits them.

If you sign up for one of the special cooking events, you will also get a glimpse of Monticello's gracious home-like interior. Among the events being considered by Monticello's culinary director, Richard Alexei, are component food and wine tastings which will focus on pairing the two.

Monticello now reserves one-third of its grapes for its own product, in addition to purchasing grapes from other vineyards for this purpose. Its current output is 18,000 cases, mostly of vintage-dated varietals, but a small amount of vin blanc and claret is also produced. The winery is most noted for its Chardonnay, Sauvignon Blanc, and Reserve Cabernet.

Open all year, daily except holidays: 10 - 4. Guided tours daily, at 10:30 and 2:30, 30 minutes. Appointments necessary (one week's notice). Free tasting. Retail outlet: wine, gifts. Special events. Access for the handicapped. Directions: Highway 29 north from Napa. Turn east on Oak Knoll Avenue. At Big Ranch Road, proceed south ¼ mile to Monticello Cellars. Mailing address: P.O. Box 2500, Yountville, California 94599.

Stag's Leap Wine Cellars
5766 Silverado Trail
Napa, California 94558
Telephone: (707) 944-2020
Owner: Warren Winiarski

Stag's Leap Wine Cellars, not to be confused with its neighbor, Stags' Leap (note the subtle difference in the placement of apostrophes), is a legendary place for a variety of reasons. Its 1973 vintage Cabernet took first prize in the Paris blind tasting of 1976, finishing ahead of Chateau Mouton-Rothschild, Chateau Haut-Brion, and Chateau Montrose, thus bringing world attention to California wines. The winery takes its name from a stag sighted on the property.

Stag's Leap is a small winery—yearly output: 25,000 cases—with an emphasis on handcrafted, premium varietal wines, particularly Cabernet Sauvignon, Merlot, and Chardonnay. Tours of the winery are by appointment only and are tailored to the interests of the participants. Visitors are usually shown the basic winery components and are encouraged to ask questions. Because of limited production, there are no free tastings. But, if you call ahead, for $1 per person plus the cost of the wine, the staff will provide a private tasting, complete with fruit and cheese—outdoors, around a wrought-iron table and chairs shaded by lofty oak trees, weather permitting. The retail sales area, occupying a corner near the working winery entrance, is open daily.

Open all year, daily except major holidays: 10-4. Guided tours, by appointment only, at 11, 1, and 3. No free tastings. For a charge, private tastings can be arranged. Retail outlet: wine, gifts. Access for the handicapped. Directions: North of Napa on Silverado Trail. Winery on east side of road.

Trefethen Vineyards
1160 Oak Knoll Drive (P.O. Box 2460)
Napa, California 94558
Telephone: (707) 255-7700
Owners: The Trefethen Family

One of the oldest wooden wineries in the Napa Valley. Trefethen's three-story structure was built in the late 1800s by Captain Hamden McIntyre, the leading architect of wineries at that time. (Captain McIntyre also designed Inglenook Vineyard's original building and the monumental Greystone Cellars.) Surrounded by 600 acres of vineyards. Famous for its Chardonnay. Open all year, Monday - Saturday: 10 - 4. Guided tours, by appointment only (one week's notice necessary) at 10:30 and 2:30. Free tasting. Retail outlet: wine. Access for the handicapped. Directions: Midway between Napa and Yountville on east side of Highway 29. (Orange-colored building is visible from the road amid a grove of trees.)

Oakville

Robert Mondavi Winery
7801 St. Helena Highway (P.O. Box 106)
Oakville, California 94562
Telephone: (707) 963-9611
Owners: The Mondavi Family

The Robert Mondavi Winery is a Napa Valley showplace. It is the oldest of the new wineries, founded in 1966 by Robert Mondavi, who with his older son, Michael, left the family business, the Charles Krug winery up the road in St. Helena, to start his own winery. The senior Mondavi, a leading figure in modern California viticulture, is known for blending tradition with an innovative approach to winemaking. His handsome facility, with its dramatic Hacienda-style lines, is also one of the largest in the valley and has excellent visitor facilities.

The winery offers one of the most comprehensive tours in the Northern Coast, a tour that leaves every ten minutes from under the arch near the visitor reception area. From there you are led into the surrounding vineyards by a well-trained guide with all the facts at his or her fingertips. After being offered a brief history of the winery and an overview of viticulture, you move on to the scale and crushing area and into the working winery, where you follow the process of making grapes into wine step by step.

After the tour, a fairly structured tasting is conducted by a guide in a small tasting room located near the retail sales area. (Tastings can only be had here with tours.) The retail sales/visitor reception room offers a good selection of wine books for sale as well as Mondavi vintages.

In addition to the regular tour, just described, the winery also offers in-depth "Harvest Tours" that last three to four hours and cover every aspect of viticulture, oenology, and the sensory appreciation of wine in great detail. Here you visit at least two different vineyards and are given a more extensive tour of the winery where fermenting and aging wines are tasted from the barrel. The tour is usually capped by a vertical tasting of Robert Mondavi wines. Harvest Tours are limited to twelve persons each and are currently scheduled for Monday, Wednesday, and Friday at 10. Appointments are required, with at least one day's notice needed. Participants are advised to wear jeans, rugged shoes or boots, and to pack a lunch.

The winery also hosts a variety of cultural and special events, some of which are free to the public. These include art exhibitions mounted in the Vineyard Room; summer jazz concerts held in the inner courtyard; and "The Great Chefs of France and America Cooking Schools," week-long seminars that in the past have featured such cooking luminaries as Julia Child, Paul Bocuse, Paul Prudhomme, Larry Forgione, and Alice Waters.

The annual output of the winery, 500,000 cases of premium varietal table wines, includes Cabernet Sauvignon, Fumé Blanc, Pinot Noir, Chardonnay, and Moscato d'Oro. Mondavi is especially known for its reds. For a very special occasion, you may want to try Opus One, a wine Robert Mondavi is producing jointly with Baron Philippe de Rothschild.

Open all year, daily except major holidays: 9-5 (Winter hours: November 1–April 30, 10-4:30.) Guided tours, every 10 minutes from the visitors' center, 45 minutes. Harvest Tour, guided, by appointment only, Monday, Wednesday, and Friday, 3 - 4 hours. Free tasting with tours only. Retail store: wine, books, and gifts. Changing art exhibits, summer jazz concerts, and cooking classes. Access for the handicapped. Directions: From Napa, Highway 29 north to Oakville. On west side of road in Oakville.

Rutherford

Beaulieu Vineyard
1960 St. Helena Highway
Rutherford, California 94573
Telephone: (707) 963-2411
Owner: Heublein, Inc.

If the *Guide Michelin* gave stars for wineries as it does for restaurants, Beaulieu would undoubtedly receive its highest rating. The winery is one of the oldest and most distinguished in the United States. Its wines have been served at the White House through nine presidential administrations. Its Private Reserve Cabernet Sauvignon, identified in February, 1980 as one of the twelve great wines of the world by Gault-Millau, the distinguished wine and food evaluators of Paris, was the only American wine named among the group. And its winemaker was none other than the famous André Tchelistcheff.

Beaulieu hugs Highway 101 in downtown Rutherford across from Inglenook Vineyards, Heublein's other holding in the Napa Valley. Its name in French means "beautiful place," a quality Rutherford probably reflected more when the winery was founded in 1900 by Georges de Latour than it does today. M. de Latour had come to California from his native France to seek gold. And when mining didn't pan out, he turned to his family heritage, winemaking, to seek his fortune. At the turn of the century he bought property in the Napa Valley and then went to France to import cuttings of the best French varieties with which to plant his vineyard. In 1923, he bought the Seneca Ewer winery across the road from his home vineyard, where Beaulieu winery is located to this day. In 1938, when his winemaker died, de Latour went to Europe to hire a successor and brought back André Tchelistcheff, a man whose expertise has changed the course of winemaking in California and in the states of Washington and Oregon, as well.

Today, the thirty-minute guided tour through Beaulieu's hallowed cellars covers crushing through bottling with a close look at how its Cabernet and other reds are produced; also included is a fifteen-minute audio-visual presentation on winemaking in the winery's formal theater. In the tasting room three to four wines can be sampled daily.

Beaulieu is famous for its Cabernet Sauvignon, but according to Leon Adams, author of *The Wines of America* (1985), better buys are its second-grade Cabernet and Burgundy, priced at about a third of the Private Reserve. Beaulieu also produces a full line of vintage varietals and generic table wines, two champagnes, and numerous dessert wines.

Open all year: daily 10-4. Guided tours, hourly: 10-3 with film (45 minutes).
Retail outlet: wine. Two small restaurants adjoining, but not managed by
BV. Access for the handicapped. Directions: Twelve miles north of Napa on
east side of Highway 29 in Rutherford.

Cakebread Cellars
8300 St. Helena Highway
Rutherford, California 94573
Telephone: (707) 963-5221
Owners: Jack and Dolores Cakebread

Cakebread Cellars is a small premium winery located just outside Rutherford on the east side of Highway 29 (across from Grgich Hills). Surrounded
by a planting of Sauvignon Blanc, the rustic redwood-sided winery offers
tours by appointment only. A small retail sales (wine only) and tasting
area, located in front of the working winery, is open daily. All wine for
sale can be sampled. Tours cover the award-winning winery and vineyards,
field crushing, fermentation, cooperage, and more. The wines for which
Cakebread Cellars is particularly noted are Cabernet Sauvignon, Chardonnay, and Sauvignon Blanc.

Open all year: daily 10 - 4. Guided tours by appointment only (3 days notice
requested). Free tasting. Retail store: wine. Access for the handicapped. Direc-
tions: Between Oakville and Rutherford on east side of Highway 29. Mail-
ing address: P.O. Box 216, Rutherford, California 94573.

Franciscan Vineyards
1178 Galleron Road (P.O. Box 407)
Rutherford, California 94573
Telephone: (707) 963-7111
Owner: Peter Eckes Company

A medium-sized, fairly new winery—built in 1973—with a tasting room
of grand proportions. Lecture on "sensory evaluation of wine" daily at 2.
View of the vineyards from mezzanine, antique corkscrew collection, and
outdoor museum. Open all year: daily 10 - 5. Self-guided tour, with signs.
Free tasting. Retail outlet: wine. Directions: From Napa take Highway 29
north to Rutherford. Franciscan Vineyards on east side just beyond Ruther-
ford at intersection of 29 and Galleron Road.

Grgich Hills Cellar
1829 St. Helena Highway (P.O. Box 450)
Rutherford, California 94573
Telephone: (707) 963-2784
Owners: Miljenko Grgich and Austin Hills

Grgich Hills derives its name not from a place, but rather from the surnames of its partners, Miljenko Grgich and Austin Hills. The two joined forces in 1977 to found the small (25,000 case) premium winery which is located just north of Rutherford on Highway 29.

Miljenko Grgich was born in Croatia-Yugoslavia, where his father was a vineyard owner. "I was stomping my first grapes when I was three years old," he recalls. Before coming to the United States in 1958, Mr. Grgich studied oenology and viticulture at the University of Zagreb. He has worked at several California wineries, including Chateau Montelena, where his 1973 Chardonnay captured first prize at the famous French-California tasting in Paris in 1976. (Another Chardonnay fashioned a few years later at Grgich Hills swept the "Chardonnay Shootout" held in Chicago in 1980 in which more than 200 Chardonnays world-wide were judged.) Austin Hills, of the Hills Brothers Coffee family, provided the backing for the Grgich Hills venture and grows the grapes. (The vineyards extend beyond the winery.)

Housed in an ivy-covered building, the winery reflects California's Spanish Colonial heritage. Its red tile roof, stuccoed walls, and arched entries are traditionally handsome. The tours, guided by either the winemaker or his assistant, cover crushing, cooperage, and bottling, and conclude with tasting.

Grgich Hills' stellar attraction is its Chardonnay. (This varietal accounts for at least half of the winery's production.) Other wines produced are Fumé Blanc, Johannisberg Riesling, and Zinfandel.

Open all year: daily 9:30 - 4:30. Guided tours: by appointment only, at 11 and 2:30 daily, 30 - 45 minutes (2 - 3 days notice needed.) Free tasting. Retail store: wine. Access for handicapped. Directions: From Napa, Highway 29 north to Rutherford. Grgich Hills on left after Rutherford Road.

Inglenook Vineyards
1991 St. Helena Highway (P.O. Box 402)
Rutherford, California 94573

Telephone: (707) 963-9411
Owner: Heublein, Inc.

Inglenook is a Scottish term meaning "warm and cozy corner," and it was this ambiance that attracted the winery's founder, Gustave Neibaum, to the Rutherford foothills of the Napa Valley in 1879. Here, at the age of 37, the Finnish sea captain, who had made a fortune in the fur trade, founded Inglenook. Building what was then the largest and most modern winery in California, he devoted the remainder of his life to the goal of producing the finest wines, wines that would "equal and excel the most famous vintages of Europe." And he evidently met with success. In 1889 Inglenook wines gained recognition at the Paris Exposition where they were awarded a coveted "General Excellence and Purity" diploma.

Captain Neibaum's legacy is still apparent when one visits the winery today, although the ownership is no longer in family hands. The original winery building is still standing in all its Victorian splendor. It now serves as the visitor reception center. Here the tours, which are historical in orientation, begin. Among the highlights is Captain Neibaum's tasting room, a replica of the captain's cabin on his ship. It is finished with carved oak paneling, stained-glass windows, and fine antiques dating to the 16th century. Other tour sites include the wine library, a collection of rare and vintage wines, and the original stone aging cellar which still retains some of its large German oak casks. Steeped as it is in history, the Inglenook tour hardly ignores modern winemaking. An eight-minute video features crushing, and in nice weather your tour group will visit the surrounding vineyards for a discussion of viticultural practices.

At the conclusion of the tour, guests are taken to a separate tasting area where white linen, fine crystal, and flowers are used to complement a particularly educational tasting. The tasting explains the uses of one's senses to evaluate wine and wine service and demonstrates how to judge wine presented in a restaurant.

Inglenook employs a culinary director and frequently hosts dinners to benefit charities and to celebrate festive holidays such as Valentine's Day. Many of these *prix fixe* dinners are open to the public. They are quite popular and pair creative cuisine with, of course, Inglenook wines.

Inglenook has an extensive product line, at the top of which are its Cabernet Sauvignon, Charbono, Pinot Noir, Zinfandel, Chardonnay, and Sauvignon Blanc varietals.

Open all year, daily except major holidays: 10 - 5. Guided tours: 10 - 4:30, 45 minutes. Free tasting with or without tour. Retail outlet: wine, gifts. Special dinners, by reservation. Directions: From Napa, Highway 29 north to Rutherford. Inglenook Vineyards on west side of highway.

St. Helena

Beringer Vineyards
2000 Main Street (P.O. Box 111)
St. Helena, California 94574
Telephone: (707) 963-7115
Owner: Nestlé, Inc.

The beautifully restored seventeen-room mansion, known as the Rhine House, that Frederick Beringer built in 1883, now serves as the visitors' reception area for the large winery that bears his family name. The winery faces Highway 29 on the north side of St. Helena, in a residential area where the house looks completely at home. The Beringer winery is a major attraction in the Napa Valley wine country and is listed on both the California and National Registers of Historic Places.

The tour here is largely historical, with emphasis on the family's achievement and on the background of the Napa Valley. (Beringer's modern working facility itself is actually across the road on the other side of Highway 29.) The guide will regale you with entertaining details about the Beringer brothers, Jacob and Frederick, who founded the winery (don't miss the stained-glass entry panels in which they depicted themselves as Shakespearean knights), and then you'll be escorted through the original winery built of native stone, which abuts the carved-out hillside behind the house. It is now used only for storage, but the sandstone tunnels carved by Chinese laborers in the late 1800s and old wine cellars dating from the '30s and '40s are still visible. The latter were discovered hidden behind enormous vats. As the tour moves outside, you'll hear the amusing story of the "soaked" oak—a massive oak tree with a taste for brandy—which, growing at a decided angle, appears to be somewhat pixilated. Nearby is a small vineyard planted to Cabernet Sauvignon to illustrate the guide's comments on grape cultivation. (If you visit in September or October when the grapes are ripe, you'll be invited to harvest and taste the fruit.) Your group will then enter the gracious Rhine House for tasting at Beringer's lovely carved tasting bar, a gracious amenity complemented by a tempting selection of fine crystal and an unusually complete collection of wine books, all of which are for sale.

Usually four wines are available for tasting at no charge, but, during the heaviest tourist season, Beringer opens its reserve wine bar in an adjacent alcove. Here, for $2 a glass, the wine connoisseur can sample reserve and older vintages.

Beringer holdings include 2000 acres in the Napa and Knights valleys

planted to Chardonnay, Sauvignon Blanc, Cabernet, Zinfandel, Riesling, Chenin Blanc, French Colombard, Gamay, and Semillon. The firm produces mainly premium varietal wines, but its product line also includes two generics—a Chablis and a Burgundy—and two dessert wines: Malvasia Bianca and Cabernet Sauvignon Port.

Open all year, daily except major holidays: 9:30 - 3:45. Guided tour, 45 minutes. Free tasting. Retail store: wine, books, gifts. Access for the handicapped. Directions: On the left side of Highway 29 on the northern edge of St. Helena.

Freemark Abbey Winery
3022 St. Helena Highway North (P.O. Box 410)
St. Helena, California 94574
Telephone: (707) 963-9694
Managing Partners: Charles Carpy and William Jaeger, Jr.

Set slightly off the road in a courtyard of buildings of faintly Tudor design is Freemark Abbey, a handsome complex consisting of the well-known Napa Valley winery, a restaurant called the Abbey (specializing in mesquite-grilled fare), a coffee garden and deli, an extensive gift shop, and a candle factory.

After you pass over the bridge/walkway from the parking lot, you'll begin your tour at the second-story tasting room which is handsomely decorated with antiques. A comfortable sofa, chairs, and piano are grouped around a massive hearth where a fire often blazes on cool winter days. The tour opens with a brief history of the winery. Then you'll move downstairs to the working facility, stopping at each step in the wine pro-

duction process along the way. You'll see the scale on which the grapes are weighed, the crushing station, fermenting tanks, barrel aging of the wine in French oak, bottling equipment, and much more. At the conclusion of the tour, which takes about twenty minutes, your group will return to the tasting room where two wines are available for sampling each day, a red and a white.

Freemark Abbey has had a long and checkered history in California. Its first grape vines were planted by Captain William J. Sayward upon his retirement from the sea. He had acquired the land in 1867 from Charles Krug, the father of the Napa Valley wine industry. The vineyards were then sold in 1881 to John and Josephine Tychson, Danish immigrants. John Tychson died a few years later, but his wife, in an unusually liberated move for a woman of that era, went ahead with their plans to build a winery, the only winery, as it turned out, to be built by a woman in California in the 1800s. The grape disease phylloxera forced Joesphine Tychson to sell the winery a few years later. Anton Forni, its next owner, enlarged the winery building and cleverly sold his wine to Italian stone masons working the marble and granite quarries in Barre, Vermont, instead of to the then glutted California market.

After a succession of owners, the property was bought in 1939 by Albert

Ahern who restored the winery to full production, changing its focus to high-quality premium wines, and naming the facility Freemark Abbey. In 1967 the winery was bought by its current owners, who completely updated it and purchased 130 acres of land in nearby Rutherford which they planted to Cabernet Sauvignon, Chardonnay, Merlot, and Johannisberg Riesling. Known as Red Barn Ranch, these vineyards provide the winery with part of the grapes required to produce its 28,000 cases annually.

Freemark Abbey produces only vintage-dated table wine. It is particularly known for its Chardonnay and Cabernet Sauvignon, and is especially proud of its Edelwein, a Late Harvest Johannisberg Riesling which has been infected by botrytis cinerea, a mold that attacks the ripened grapes and results in a high concentration of sugar.

Open all year, daily except Thanksgiving, Easter, Christmas, and New Year's: 10 - 4:30. Guided tour at 2, approximately 20 minutes. Free tasting. Retail store: wine and gifts. Access for the handicapped. Directions: ½ mile north of St. Helena on Route 29. Freemark Abbey on right.

Greystone Cellars
2555 Main Street
St. Helena, California 94574
Telephone: (707) 226-5566
Owner: Brothers of The Christian Schools

Unfortunately, at present, you can only see the outside of the monumental Greystone Cellars, perhaps the most impressive old stone cellar in all the Napa Valley. (It is located on the left as you travel north from St. Helena on Route 29.) The winery is temporarily closed for structural strengthening needed because of shifting earth, but word is that it may reopen soon. In the interim, tasting and retail sales take place out front in a circus tent. The Christian Brothers (actually the Brothers of the Christian Schools, the Catholic Church's largest order of men dedicated solely to teaching) were among the first major vintners to settle in the Napa Valley fifty years ago and have the largest holding of premium vineyards in the area. A storage and aging capacity of eight million gallons and the largest stock of aged premium wines in the valley attest to this modern tax-paying corporation's size and greatness. There's lots to see, learn, and taste here when Greystone Cellars reopens, including Brother Timothy's extensive corkscrew collection. Call or write for an update on opening status. Mailing address: P.O. Box 420, Napa, California 94559.

Heitz Wine Cellars
436 St. Helena Highway South
St. Helena, California 94574
Telephone: (707) 963-3542
Owners: The Heitz Family

The primary location of the Heitz vineyards and winery is the family
ranch on Taplin Road in Spring Valley, east of the town of St. Helena.
Tours of the ranch/winery are by appointment only, since the Heitzes
also live there. But if you are out for the day touring the Napa Valley
wineries, you can sample Heitz wines by stopping at the family's tasting
and sales room conveniently located on Route 29 just south of St. Helena.
This was the site of the original Heitz winery, and the small redwood-
sided structure still provides supplementary aging space.

Heitz is especially known for its Martha's Vineyard Cabernet Sauvignon
and its Chardonnays.

*Tasting (no charge) and sales room open daily, 11:00 - 4:30. Access for the
handicapped. Directions: Highway 29, 1½ miles south of St. Helena. On east
side of road. Mailing address: 500 Taplin Road, St. Helena, California 94574.*

Charles Krug Winery
2800 Main Street (P.O. Box 191)
St. Helena, California 94574
Telephone: (707) 963-2761
Owners: The Mondavi Family

*An old Napa estate winery with massive stone buildings. Founded in 1861
by Charles Krug and now owned by the Mondavi Family. Very complete
tour of grounds. Open all year: daily 10-4:45. Guided tours, between 10 and
4. Free tasting. Retail store. Outdoor concerts in August. Directions: east
side of Highway 29 on northern edge of St. Helena.*

Louis M. Martini
254 South St. Helena Highway (P.O. Box 112)
St. Helena, California 94574
Telephone: (707) 963-2736
Owners: The Martini Family

Louis M. Martini, the son of an Italian immigrant who ran a fish business, founded the winery that bears his name more than fifty years ago. Known as the "grand old man" before his death at age 87 in 1974, Mr. Martini became something of a legend in California for both his innovative approach to winemaking and as a leader of the emergent premium wine industry in the state following the repeal of Prohibition.

Today the Louis M. Martini winery—which has more than doubled its capacity since its founder's time to 3.5 million gallons—continues to be a family-owned-and-operated business. A tour of this "no frills" winery is particularly interesting for the innovative equipment the elder Martini developed—much of which is still in use, such as the open concrete fermenters used for the red wines and the huge refrigerated room of redwood tanks used for fermenting white wines before refrigerated stainless-steel tanks came into use. Afterwards a visit to the tasting room reveals a generous attitude towards sampling, with all of Martini's regular vintage varietal wines generally available.

The winery's product line includes five vineyard selection wines, special selection wines, vintage varietal wines, aperitif and dessert wines, and generic wines. The winery is especially known for its policy of reasonably pricing its aged wines.

Open all year, daily except major holidays: 10 - 4:30. Guided tours, regularly scheduled between 10 and 3:30. Free tasting. Retail outlet: wine. Access for the handicapped. Directions: Winery located on left side of Highway 29, ½ mile south of St. Helena.

Newton Vineyard
2555 Madrona Avenue
St. Helena, California 94574
Telephone: (707) 963-4613
Owners: Richard Forman; Peter and Su Hua Newton

An up-and-coming winery with 100 acres of dramatically terraced vineyards. (Peter Newton was the principal founder of Calistoga's Sterling Winery; Forman, its winemaker.) Newton Vineyard is located outside of St. Helena, near Spring Mountain Vineyard. Formal boxwood gardens hide the roof of an in-ground aging cellar and an observation tower provides views of the valley. Open all year, by appointment only: Monday - Friday, 9 - 5. Guided tours. Retail sales. Directions: Highway 29 north to St. Helena. Left onto Madrona Avenue/Spring Mountain Road, beyond town center. Winery on west side of road next to Spring Mountain.

Joseph Phelps Vineyards
200 Taplin Road (P.O. Box 1031)
St. Helena, California 94574
Telephone: (707) 963-2745
Owner: Joseph Phelps

Building wineries for others (such as the impressive Souverain facility in Sonoma County) obviously whetted Joseph Phelps's appetite for building one for himself, which is just what the originally-Colorado-based builder did in the early '70s on a 625-acre tract of land, long known as the Connolly ranch, east of St. Helena, on Taplin Road.

The winery, offices, and laboratory of Joseph Phelps Vineyards are housed in a most dramatic piece of architecture which integrates large rough-sawn redwood planks in its design, as well as recycled timbers from a railroad trestle. (The latter material is also used imaginatively in the design of the entry gate.) The vineyards are planted on only 175 acres of the large property, allowing for careful matching of grape variety to soil and growing conditions.

Here Cabernet Sauvingon and Zinfandel face the afternoon sun; Gewürztraminer and White Riesling grow in a low dip abutting the Napa River, where the risk of spring frost is offset; and one of California's few plantings of French Syrah occupies several acres. The remaining acreage is largely natural, thus "preserving the peaceable nature of Spring Valley," a stated aim of the winery. Phelps Vineyards also owns additional acreage in the cooler Yountville area. This land is devoted mainly to Johannisberg Riesling, Gewürztraminer, and Chardonnay. In addition, the winery also purchases grapes from selected ranches, the finest of which (Eisele and Bakus, for example) are frequently identified on its labels.

The dramatic wooden cellars built on a west-facing slope of the property consist of two pavilions joined by a closed bridge. The northernmost pavilion contains the fermentors, as well as small upright oak tanks and oak oval casks from Germany. The other pavilion holds lofty racks full of oak barrels from France and the bottling line. Between the pavilions, the enclosed bridge houses offices and a laboratory. Guided tours include these facilities, a view of the vineyards, and a tasting.

Joseph Phelps Vineyards is known for its varietal table wines, particularly its Riesling, Cabernet Sauvignon, Chardonnay, and Syrah. (It is one of the few wineries to own twenty acres of the true French Syrah grape, a grape not to be confused with the more widely grown Petit Sirah.) Phelps also produces inexpensive quality generics called vin rouge and vin blanc. The firm has won numerous awards for its wines, among them a gold medal from the American Beverage Testing Institute for its 1981 Chardonnay "Schellville."

Open all year: Monday - Friday, 9 - 4; Saturdays 10 - 4. Guided tours, by appointment only (24 hours notice needed): Monday - Friday at 11 and 2:30; Saturdays at 10, 11:30, 1, and 2:30. Free tasting only with tour. Retail outlet: wine, books, and gifts. Picnic tables available by reservation. Access for the handicapped. Directions: Highway 29 north to St. Helena. Right turn at Zinfandel Lane to Silverado Trail. Left on Silverado Trail to Taplin Lane. Right on Taplin to the winery.

Spring Mountain Vineyard
2805 Spring Mountain Road
St. Helena, California 94574
Telephone: (707) 963-5233
Owner: Michael Robbins

If you don't care to pay the $3 fee for touring the grounds, you can always see Spring Mountain's splendid Victorian mansion, vintage 1880, from the road. The fee was imposed to keep away at least a few of the hoards of tourists that come here more to see the site of CBS-TV's prime-time "soap" *Falcon Crest* than for the wine!

Falcon Crest notwithstanding, the property was well known in an earlier time as Villa Miravalle, the home of Tiburcio Parrott, a wealthy and eccentric bachelor who was a leader in experimental agriculture in the Napa Valley in the 1800s. Parrott had a reputation for trying to raise almost everything. His horticultural repertoire ranged from olive trees and palmettos to persimmons, pomegranates, tobacco, and grapes, and some of these plantings still remain on the property today. But he was most successful with his vineyards, which were principally planted to Cabernet Sauvignon, and for the wine he made from this varietal. Before the days of Prohibition, the greatest California wine was considered to be that of Miravalle.

Following Parrott's death in 1894, the villa remained vacant for more than forty years. It then housed a roller-skating rink among other less than distinguished tenants, until in 1974 its next full-time resident and current owner, Michael Robbins, fully restored the property, including one of the 19th-century caves which still remained intact. Mr. Robbins also built a new winery consistent with the architecture of the original villa.

Today Spring Mountain's annual production is 25,000 cases of table wine. Of the 265 acres surrounding the winery, 35 are currently planted to grapes, with 65 more to be planted soon. Grape varieties grown are Chardonnay, Sauvignon Blanc, Cabernet Sauvignon, Merlot, Pinot Noir, and Cabernet Franc. Spring Mountain is most noted for the wines made from its Cabernet Sauvignon and Chardonnay.

Open all year: daily 10 - 4. Guided tours, by appointment (make these as far ahead as possible since weekend tours are often booked long in advance): Monday - Friday at 10:30 and 2:30, and Saturday and Sunday at 10:30, 30 minutes. Free tasting. Historic walking tour of the estate: 11 - 4 daily, $3 adults, $1 children 7 - 12. Access for the handicapped limited to the tasting room. Directions: North on Highway 29 to St. Helena. At traffic light in town take left to Madrona, right on Spring Mountain Road. One mile on Spring Mountain to winery.

Yountville

Domaine Chandon
California Drive
Yountville, California 94599
Telephone: (707) 944-8844
Owner: Moët-Hennessy

The ultimate in understated elegance in the Napa Valley is the Moët-Hennessy holding, Domaine Chandon in Yountville. Set apart from the heavy concentration of wineries which line Highway 29 near St. Helena, its modern buildings combine design elements from both California and Champagne, France. (The native stone and exposed beams are characteristic of the Napa area, while the arched doorways and roof lines were inspired by the caves of Champagne.)

Hidden from the road by a gentle slope of land, its museum-like architecture is revealed on the right as you approach the winery along its entrance road. The winery buildings are handsomely framed in the foreground by a naturally landscaped pond with waterfall, reeds, and rocks, and in the distance by the Mayacamas Mountains. A footbridge over the stream which feeds the pond leads to the gallery where visitors assemble for the tour. This area, lined with changing exhibits (on a recent visit the cooper's art and the glassmaker's craft were featured), illustrates different aspects of the Champagne-making process. Usually led by a French intern, the tours are quite technical. (Domaine Chandon is an excellent place to start if you want to learn about the ins

APÉRITIF
WINE

ALCOHOL
18% BY VOLUME

PRODUCED
AND BOTTLED
BY DOMAINE
CHANDON
YOUNTVILLE
CALIFORNIA

SERVE CHILLED
OR OVER ICE

SERVIR FRAIS

and outs of the making of sparkling wine. The facility was obviously designed to place great importance on visitor education and hospitality.)

Starting with the well mounted "museum" exhibits—which include a collection of antique vineyard and winery tools from Champagne and two large 19th-century presses—the tour moves on to the making of sparkling wine today in the working winery. All aspects of sparkling wine production are covered in the 30 to 45 minute tour: from grape-growing techniques to primary fermentation, bottle fermentation and aging, riddling, disgorging, and dosage. (The disgorging line where the sediment is removed, the dosage added, and final corking accomplished is not in operation on Friday, Saturday, or Sunday, so if you want to see these steps, plan your visit accordingly.)

After the tour, visitors are invited to the charming tasting salon, where Domaine Chandon's two sparkling wines—Napa Valley Brut and Blanc de Noirs—and a Pinot Noir-based apéritif—Panache—can be sampled. There is a charge of $2.25 for each flute, but this is not an uncommon policy with sparkling wine because of the expense involved in its production. The tasting salon, one should add, is particularly attractive. Colorful chintz covers the tables, complimentary hors d'oeuvres are provided, and tour guides generally serve their groups and are available to answer any questions about the product.

Domaine Chandon also has a four-star restaurant, serving both lunch and dinner, adjacent to the salon. Designed to showcase its wines and the regional cuisine of Champagne, it is one of the best, and deservedly popular, restaurants in the Valley. Reservations are essential, especially in the busy summer and fall.

During the summer season, Domaine Chandon offers free concerts which are open to the general public. There are usually two concert series presented on the terrace in June and September. And special entertainment is also planned for July 4th and July 14th (Bastille Day).

Open daily: May - October, 11 - 5:30. (Closed Mondays and Tuesdays, November - April.) Guided tours, regularly scheduled, leave on the hour weekdays, and on the half hour weekends (30 - 45 minutes). Charge for tasting. Retail area: wine, books. Four-star restaurant for lunch and dinner. Reservations: (707) 944-2892. Free summer concerts. Access for the handicapped. Directions: Nine miles north of Napa, exit Highway 29 at Yountville Veterans Home. At end of ramp, turn west toward Veterans Home, cross railroad tracks and turn right into Domaine Chandon.

SONOMA VALLEY AND SONOMA COUNTY

Asti

Italian Swiss Colony
26150 Asti Road (P.O. Box 1)
Asti, California 95413
Telephone: (707) 433-2333, (707) 894-2541
Owner: ISC Wines

Nostalgic spot for a generation that quenched its thirst on rafia-covered jugs of chianti and then recycled the "empties" into candelabras. The winery's "little old winemaker" ad campaign made Swiss Colony a household word. Large Swiss-chalet-style tasting room. Famous church—shaped like a wine

*barrel—on adjoining property. Open all year, except major holidays: 10-5.
Guided tours, regularly scheduled throughout the day, 30 minutes. Free tasting
(with or without tour), usually 5 to 7 wines available daily. Retail outlet:
wine, gifts, deli. Extensive picnic facilities. Access for the handicapped. Directions: Highway 101 north from Healdsburg. At Asti exit, take a right; winery
straight ahead, 300 yards. (Colony is the town of Asti; the Asti branch of
the U.S. Post Office is even located in the winery buildings.)*

Geyserville

Geyser Peak Winery
22280 Geyserville Road (P.O. Box 25)
Geyserville, California 95441
Telephone: (707) 433-6585
Owners: The Henry Trione Family

The large (2.5 million gallons) Geyser Peak Winery, located in northern
Sonoma County, began its life as a producer primarily of bulk wines and
brandy for other wineries. In 1972, when it was bought by the Joseph
Schlitz Brewing Company, its focus changed to table wines. (The Summit label identifies the winery's jug wines, while the Geyser Peak label
is reserved for vintage-dated varietals—the latter representing about 200,000
cases annually.)

The winery's original fieldstone and wood structure, built by August
Quitzow in 1880, is still standing. But it is eclipsed by the facility's centerpiece, the grand chateau-like tasting room/visitors' center added during
the Schlitz era.

There are plenty of picnic tables. (One group is located next to the
original winery building, the other in a nearby vineyard.) After lunch
there are two hiking trails to explore. One leads along the Russian River;
the other, a short trek up the hill behind the winery, affords a view of
the valley. (There are some picnic tables at the end of this trail as well.)

Under ownership of the Triones, a Santa Rosa grape-growing and bank-
ing family that purchased Geyser Peak in 1982, the winery has added two
methode champenoise sparkling wines, two soft (8.5 percent) wines, and
some estate-bottled vintage wines to its product line.

*Open all year: daily 10-5. Free tasting (usually eight wines available). Retail
outlet: wine, gifts, and nonalcoholic grape juice. Extensive picnic facilities.
Hiking trails. Summer operas in the vineyards. Access for the handicapped.
Directions: Highway 101 north from Healdsburg to Canyon Road exit. Right
at exit. Left under freeway; winery 1/10th mile beyond on your right.*

J. Pedroncelli Winery
1220 Canyon Road
Geyserville, California 95441
Telephone: (707) 857-3531
Owners: John and Jim Pedroncelli

Family-owned-and-operated winery a short distance west of Geyser Peak Winery on Canyon Road. Run by second generation of Pedroncellis: brothers John and Jim. Redwood structure and masonry additions hold working winery and tasting room. Vineyards surround. J. Pedroncelli was one of the first California wineries to shift from bulk wine production to premium varietal wines. Noted for Vintage Selection Cabernet and Zinfandel. Wines associated historically with good value for the price, and winery famed for producing California's first Zinfandel Rosé to achieve commercial fame. Open all year: daily 10-5. Free tasting. Retail outlet: wine. Access for the handicapped. Directions: Take Highway 101 north from Healdsburg to Canyon Road exit. At exit, go west on Canyon Road approximately one mile to winery.

Souverain Cellars
Independence Lane (P.O. Box 528)
Geyserville, California 95441
Telephone: (707) 433-8281
Owner: North Coast Cellars

The most dramatic of the Geyserville wineries, and visible from Highway 101, is Souverain Cellars. Built in 1972 by Pillsbury, its former owner, the winery boasts modern architecture reminiscent of the hop kilns that dotted the Sonoma Valley in days gone by. Two angular "hop kiln" towers flank either end of the winery, connecting the 320-foot-long working facility. A formal courtyard and central fountain fill the elongated "U" shape. The tower to the left (as one approaches the building from the parking lot) serves as the visitors' hospitality center. Upstairs there is a tasting area and retail outlet; downstairs, a restaurant.

Regularly scheduled guided tours are offered daily at Souverain. Here you can walk through the whole facility and observe the bottling line, lab, and fermentation area, as well as the large storage area, dominated by wood cooperage and many different-sized tanks, some as large as 30,000 gallons.

Tours conclude in the tasting room, although you don't have to participate in a tour to taste the wine. On a recent visit, twelve wines were available for sampling. The selection included both dry and semi-dry whites, blush and rosé wines, reds, and dessert wines.

Souverain's own gourmet restaurant overlooks the vineyards, specializes in regional cuisine, and, naturally, showcases Souverain's product. Lunch and dinner are served daily. And in nice weather you can dine *alfresco* on the terrace.

Several times a year the winery hosts such special events as concerts and art shows. A popular Hollyberry Crafts Faire is always held the Friday, Saturday, and Sunday following Thanksgiving. Many of these special events take place in the winery's handsome courtyard.

Souverain's product line is devoted solely to table wines. Included on its lengthy list are vintage-dated 100-percent varietal wines, vintage-dated generic wines, nonvintage table wines, and a line of estate-bottled vintage selection wines. Souverain is particularly noted for its Vintage Selection Cabernet Sauvignon, Vintage Selection Sauvignon Blanc, Colombard Blanc, and Pinot Noir Rosé.

Open all year: daily 10-5. Guided tours at 11, 1, and 3. Free tasting. Retail outlet: wine, gifts. Gourmet restaurant for lunch and dinner, daily. (Reservations suggested.) Special events. Access for the handicapped to restaurant only. Directions: South on Highway 101 from Geyserville to Independence Lane exit. Winery west of highway on Independence Lane.

Trentadue Winery
19170 Redwood Highway
Geyserville, California 95441
Telephone: (707) 433-3104
Owners: Leo and Evelyn Trentadue

Across the highway from Souverain is Trentadue, a family-owned vineyard and winery (200 acres, 25,000 cases annually) founded in 1969 by Leo and Evelyn Trentadue. The second-story tasting room, from which one can look down into the working winery below, contains a large selection of gift items and a deli. Surrounding vineyards are planted to some twenty varietals including Aleatico, Zinfandel, Nebbiolo, and Petite Sirah. The winery is noted for the wines produced from the last named. Open all year, daily except Christmas, New Year's Day, Easter, and Thanksgiving: 10–5. Guided tours by appointment only (one week's notice). Free tasting. Retail outlet: wine, gifts, deli. Picnic facilities. Directions: Three miles north of Healdsburg to Independence Road exit. East at the exit, and then north for ½ mile on Redwood Highway.

Glen Ellen

Glen Ellen Winery and Vineyards
1883 London Ranch Road
Glen Ellen, California 95442
Telephone: (707) 996-1066
Owners: The Benziger Family

Take the 5 miles per hour sign, posted near the white farm fence at the entrance to Glen Ellen, seriously. The lush route to the winery, lined with giant eucalyptus and vineyards, is an unpaved country lane, full of potholes and ruts.

Glen Ellen is a family affair in the truest sense of the term. Its day-to-day operation is managed by Bruno and Mary Benziger and five of their seven children, some with Benziger grandchildren in tow. They all live at the end of the lane in the family compound—the senior Benzigers in the landmark-designated gingerbread Victorian, circa 1867; the married children in the adjacent frame buildings. A short distance from the family's quarters and down a rather steep slope, are the combined working facility, aging cellar, tasting room, and offices. These are housed in a white barn-like structure the family built with its own hands.

There is no formal tour here, but the tasting and retail sales area is right in the middle of the working winery, so you can see what's happening and ask any questions about the winemaking process. Mary Benziger, referred to as "the mother of the winery," is usually behind the tasting area tables. Other family members are generally around, too. And everyone is very friendly.

Barrel tastings can usually be had upon request. And if your visit coincides with the busy "crush," you can even lend a hand sorting the grapes, if you wish.

On the terraced acreage surrounding Glen Ellen's home ranch, and at the family's other property, the Carnero de la Sonoma Vineyard, thrive several grape varietals: Cabernet Sauvignon, Sauvignon Blanc, Chardonnay, Cabernet Franc, Merlot, Muscat Canelli, and Semillon. Glen Ellen is particularly noted for its Cabernet Sauvignon, Sauvignon Blanc, and Chardonnay. But generic wines are also produced. All Glen Ellen wines are quite attractively priced.

Open all year: daily 10-4. Self-guided tour of winery, with family on hand for questions. Special tours of the vineyards by appointment. (One week's notice required.) Free tasting. Retail area: wine, gifts. Picnic facilities. Directions: North on Highway 12 from Sonoma. Left on Arnold Drive to London Ranch Road. Left on London Ranch to winery entrance on your right.

Guerneville

Korbel Champagne Cellars
13250 River Road
Guerneville, California 95446
Telephone: (707) 887-2294
Owners: The Adolf Heck Family

Korbel Champagne Cellars was founded in 1882 by the brothers Francis, Joseph, and Anton Korbel, who had emigrated to San Francisco from their native Bohemia. Moving north to the Russian River Valley, the brothers founded a logging business in the virgin forests, but switched to viticulture and winemaking when their land was stripped of its trees. They began making champagne in 1876. Nearly seventy-five years later, in 1954, the winery was bought by yet another trio of brothers, this time with Alsatian roots: Adolf, Paul, and Ben Heck. The Hecks and their descendants kept the Korbel family name and have continued to produce Korbel champagne in the traditional French way, making the winery the oldest producer of *methode champenoise* champagne in the United States.

To take in everything available to visitors at the Korbel winery will take a good hour and a half. Guided tours begin at the train depot—to the right of the century-old cellars. (The Korbel brothers purportedly bought the depot from the Northwestern Pacific Railroad for five dollars.) The tour group then moves into the original winery building—constructed of handmade bricks—where a fascinating exhibit of old champagne-making artifacts, tools, and photographs is on display. Next visitors see a slide show covering the history of champagne making and then proceed into the cellars with their large German ovals. Riddling—the labor-intensive process of settling the sediment in the neck of the bottle by giving it a quarter of a turn each day—is discussed, and the ingenious mechanized riddler, developed by Adolf Heck, is demonstrated. Disgorging, dosage, corking, and labeling processes are shown next, with antique apparatuses juxtaposed to their modern counterparts.

After the tour, samples of the Korbel product can be sipped in the tasting room, a refurbished brandy warehouse located to the left of the cellars. Adjoining the tasting room is the retail outlet, where wine, picnic fare, cookbooks, jams, and the like are sold. There is a picnic area on the property as well, and you are free to walk around the brick tower, located to the right of the cellars, which housed a brandy still in the Korbels' time.

Up the hill from the tasting/retail area is the circa 1880 house where the Korbels lived. Now restored, the house and its spectacular gardens can be toured by appointment. The beds of antique rosebushes are particularly attractive. Across the road from the winery are the seemingly endless Korbel vineyards, 600 acres planted to the champagne grapes: Pinot Noir and Chardonnay.

Korbel's production line includes five styles of champagne; and you are free to sample them all in the tasting room at no charge. Korbel Brut was served at President Reagan's second inauguration.

Open all year, daily except on major holidays: 10-5. Guided tours daily: 10-3. Free tasting. Retail outlet: wine, gifts, deli. Picnic area. Access for the handicapped. Directions: Take Highway 101 north from Santa Rosa to River Road exit. West on River Road exit (14 miles) to winery.

Healdsburg

Alexander Valley Vineyards
8644 Highway 128
Healdsburg, California 95448
Telephone: (707) 433-7209
Owners: The Wetzel Family

Due east of the Simi Winery and the town of Healdsburg on Highway 128 heading south towards the Napa Valley is Alexander Valley Vineyards. A small family-owned-and-operated premium winery, Alexander Valley produces 25,000 cases of estate-bottled, vintage-dated varietal wines yearly.

The winery, which was founded in 1963, is located on property settled in 1842 by Cyrus Alexander—the Alexander for whom the valley is named. The estate includes the restored 19th-century Alexander homestead and a schoolhouse built in 1868, as well as the modern Wetzel home and the working winery.

Surrounded by family-owned vineyards, the winery is a model of compactness and efficiency: It can be operated by one person, I was told, and is built against the side of a hill to take advantage of both gravity flow and the naturally cooling effect of the earth. Outdoors is the stemmer-crusher and Willmes basket press; indoors, also on ground level, are the stainless-steel fermenters; downstairs, an aging cellar is filled with American and French oak barrels; upstairs houses the small tasting room and the business offices.

Tours of the winery are by appointment only. But the airy tasting room, furnished with antiques and with windows looking out onto 140 acres

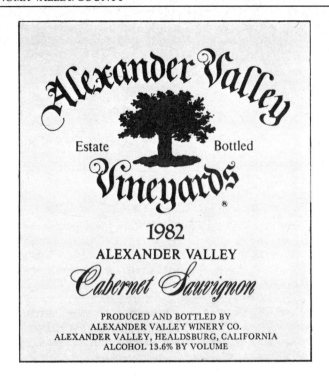

of vineyards, is open daily for tasting and sales. Wrapped around the tasting room is a balcony where you can get an even better view of the estate and enjoy a picnic lunch. You are also free to walk over and admire the Cyrus Alexander house or meander up the hill to the old grave site where Alexander and his family are buried.

Alexander Valley's product line includes Chardonnay, Chenin Blanc, Johannisberg Riesling, Gewürztraminer, Cabernet Sauvignon, Pinot Noir, Merlot, and Zinfandel. The winery is particularly noted for its Chardonnay and Cabernet.

Open all year, daily except major holidays: 12-5. Guided tours by appointment only (one day's notice). Free tasting. Retail outlet: wine. Limited picnic facilities. Directions: North on Highway 101 to Dry Creek Road exit, right to Healdsburg Avenue. Left to Alexander Valley Road. Straight to Highway 128 South. Alexander Valley Vineyards on your left.

Field Stone Winery
10075 Highway 128
Healdsburg, California 95448
Telephone: (707) 433-7266
Manager: John Staten

Plan to stop at Field Stone winery if you're visiting Alexander Valley Vineyards, for it's just down the road from the latter, and unique architecturally. Field Stone, as its name suggests, is faced with fieldstones which were culled from the excavation of this underground winery, reportedly the first built in California in this century.

Carved into a verdant knoll topped by massive oak trees, Field Stone is a small estate winery. It was founded by the late Wallace Johnson on property originally called the Montvale Ranch, which he purchased in 1955. A former mayor of Berkeley, Johnson renamed the property the Redwood Hereford Ranch, and later, after test plantings of grape varietals confirmed the land as a prime vineyard acreage, changed its name yet again to the Redwood Ranch and Vineyard. He then conceived the idea for an underground winery to crush, vinify, and bottle a select portion of his crop and completed Field Stone in 1977, just in time for the first vintage. After Johnson's untimely death two years later, the winery continued in family hands and is now managed by his son-in-law, John Staten, an ordained Presbyterian minister, in conjunction with the eminent André Tchelistcheff, consulting oenologist, and James Thomson, winemaker.

Brief tours of the unusually compact facility are occasionally offered, depending on the work load in the winery, but visitors are always free to picnic on its grassy roof, under the oak trees, with a view of the 150 acres of vineyards planted to white Riesling, Sauvignon Blanc, Cabernet Sauvignon, Petite Sirah, Chenin Blanc, and Gewürztraminer stretching into the horizon. And free tastings of the Field Stone product are offered daily.

Field Stone continues to produce 10,000 cases of varietal table wines yearly. These are all estate-bottled or vineyard-designated and vintage-dated. The winery is particularly known for its Petite Sirah and Cabernet Sauvignon.

Open all year: daily 10-5. Occasional guided tours. Free tasting. Retail outlet: wine, gifts. Picnicking permitted on grounds. Summer concerts, German festival, Fall Open House. Access for the handicapped. Directions: Located on Highway 128 halfway between Geyserville and Calistoga at junction of Chalk Hill Road.

Hop Kiln Winery at Griffin Vineyards
6050 Westside Road
Healdsburg, California 95448
Telephone: (707) 433-6491
Owner: L. Martin Griffin, M.D.

At the bend of an idyllic country road near the town of Healdsburg in the Russian River Valley, the characteristic towers of an authentic turn-of-the-century hops barn mark your arrival at Hop Kiln Winery. "Converting it to a winery was the only way to save it," says Dr. L. Martin Griffin, the winery's owner, of the wood and stone Austrian-Victorian-style barn, which has been called one of Sonoma County's most beautiful historic structures.

Originally built to dry hops, an ingredient in making beer, a once thriving industry in the area, the barn was part of the 240-acre sheep ranch that Dr. Griffin, a public health physician and conservationist, bought in 1960. "The barn was in need of structural work to preserve it from the general decay of time, but we couldn't afford it unless the building paid its way." So Dr. Griffin, who had become interested in wine while living in Italy in the early '60s, planted 65 acres of the ranch to vineyards and recycled the structure into a winery.

Hop Kiln is a charming, intimate place to visit. In his restoration, Dr. Griffin was as faithful to the integrity of the original building as possible. The exterior of the structure appears to be largely untouched. A fieldstone path to the winery winds between a few shaded picnic tables to the entrance of the tasting room/sales area, which is on the upper level of the barn. The interior is quite rustic, with exposed beams, hand-wrought hardware, and even some of the original equipment used for drying hops preserved. From the railing around the tasting area you can see the small working winery below, while a wall of new, large windows affords a view of the vineyards and the duck pond beyond. Tours are only by appointment, but you can see most everything from these two vantage points. And any questions you might have can surely be answered by the staff behind the tasting bar a few steps away.

Changing exhibitions of local artwork add a colorful and interesting note to the rather dark, weathered barn walls, as does the permanent display of vintage photographs from the barn's hops-growing era. (There are some hops planted outside if you are curious about what the plant looks like.)

Hop Kiln produces 8,000 cases of mostly still table wine annually. (One percent of its production is sparkling wine.) Sixty-five acres of the surrounding ranch are planted to Petite Sirah, Zinfandel, Early Burgundy, Napa Gamay, Johannisberg Riesling, Gewürztraminer, Chardonnay, and French Colombard. The winery is particularly known for its varietal wines—Petite Sirah, Zinfandel, Gewürztraminer—and its proprietary wines: Big Red and A Thousand Flowers.

Open all year: daily 10-5. Guided tours, by appointment only (two weeks notice suggested), 15-30 minutes. Free tasting daily. Retail outlet: wine, gifts. Art gallery. Picnic facilities. Directions: North on Highway 101 to Healdsburg exit. Left on Mill Street, which turns into Westside. Continue 6 miles south on Westside to winery.

Rodney Strong Vineyards
Piper-Sonoma
11455 Old Redwood Highway
Healdsburg, California 95498
Telephone: (707) 433-6511
Owners: *Rodney Strong Vineyards* – Renfield Importers, Inc.;
  Piper-Sonoma – joint venture of Rodney Strong, Piper-Heidsieck, and Renfield Importers.

Two wineries inspired by the same man, Rodney Strong, share a parking lot on the old Redwood Highway north of the village of Windsor. A dancer who turned his creative energies to winemaking in 1959, Strong, together with his wife, Charlotte, established his first vineyard in the town of Tiburon in Marin County. At that time the winery was called Tiburon Vintners. There the Strongs implemented a very successful marketing idea by labeling bottles of wine with the purchaser's name (viz., "bottled expressly for . . ."). As a direct result, just three years later they moved the rapidly expanding winery north to the Windsor area. Since then, the firm has grown tremendously, changing its name several times, from Windsor to Sonoma Vineyards, and most recently to Rodney Strong Vineyards. Although it is now owned by Renfield Importers Inc. of New York, Strong remains its winemaker.

Rodney Strong Vineyards' modern facility was designed as four separate wings joined together by a central working area. (The whole entity is shaped roughly like a cross.) In the core are the winemaking tools: the pumps, filters, and centrifuges. The wings are divided equally between stainless-steel fermentation tanks and oak cooperage. The combined visitors' center and tasting area is suspended above the core of the winery. Its balconies, furnished with tables and chairs, provide comfortable van-

tage points from which to view the workings below. Guided tours through the winery are scheduled on the hour and conclude with a free tasting. The tour and tasting usually take between fifteen and twenty minutes, giving you just about enough time to catch the tour at Piper-Sonoma (across the shared parking lot) which leaves on the half hour.

The tour of the glass and concrete Piper-Sonoma winery, built in 1980 as a joint venture between Rodney Strong and Piper-Heidsieck of France, takes in the latest advances in the production of *methode champenoise* sparkling wine. Here you will see mechanical riddling and a fully automated disgorging, dosage, corking, and labeling line. The facility has its own visitors' center, separate from Rodney Strong Vineyards, where for a nominal charge you can sample the Piper-Sonoma product.

Rodney Strong Vineyards is noted for its estate-grown varietal table wines, especially Alexander's Crown Cabernet Sauvignon and Chalk Hill Chardonnay. Piper-Sonoma produces only vintage-dated *methode champenoise* sparkling wines. Its product line includes Brut, Tête de Cuvée, and Blanc de Noirs.

There are extensive picnic facilities on the large green at the Rodney Strong Vineyards. This lawn, also called the "Greek Theatre", is the setting for summer music festivals featuring performances as disparate as rock and ballet.

Both Rodney Strong and Piper-Sonoma are open all year, daily 10-5. Guided tours of each facility, regularly scheduled from 11-4 daily, take 15-20 minutes. At Rodney Strong, tours leave on the hour; at Piper-Sonoma, on the half hour. Free tasting at Rodney Strong; charge for tasting at Piper-Sonoma. Retail outlets at both: wine and gifts. Picnic facilities and summer festivals at Rodney Strong. Access for the handicapped at both. Directions: North on Highway 101 from Santa Rosa. Exit west at Windsor. Take Old Redwood Highway north 3 miles to winery which faces highway on your left. Mailing address: P.O. Box 368, Windsor, California 95492.

Simi Winery
16275 Healdsburg Avenue (P.O. Box 698)
Healdsburg, California 95448
Telephone: (707) 433-6981
Owner: Moët-Hennessy

The serene expression and grandmotherly air of Isabelle Simi, captured in an old photograph on display in the hexagonal-shaped visitors' center at the Simi Winery, belies the fiber of the woman who kept the winery going, through good years and bad, for over sixty years. Located one-half mile north of Healdsburg, the winery was founded in 1876 by Giuseppe and Pietro Simi who called it Montepulciano, after the Italian village from which they had emigrated. In 1890, employing both Italian and Chinese laborers, the brothers built the original stone winery, located directly behind the present visitors' center over the railroad tracks. The two groups apparently worked independently of each other, and, consequently, the stonework on the facade of the building reflects where the two cultures met. (The Chinese style was to set the stones in a random fashion, while the Italians' masonry was far more formal.) At the brothers' untimely deaths in 1904, Isabelle, Giuseppe's daughter, then still a teenager, assumed control of the winery and eventually gave it the family name. In 1970, after more than six decades, Isabelle sold the business, which passed through several owners before its purchase in 1980 by Moët-Hennessy to complement the firm's sparkling wine holding in Napa, Domaine Chandon.

In recent years the winery has been completely renovated. The original stone building was retrofitted and now serves as an aging cellar. A new state-of-the-art fermentation facility was constructed beyond. The tour at Simi covers all of these points, both historical and physical, and follows the route of the grapes in the winery from the receiving station to the crush platform, barrel cellar, and bottling line. Visitors can taste almost the entire Simi line every day, either before, after, or without a tour. And frequently some special wines not made for the national market are also offered for their educational value (for example, a recent lot of Semillon left over from blending Sauvignon Blanc). The attractive tasting room stocks such gift items as local jams, vinegars, and wine glasses. These items are interspersed with early photos of the winery, including the portrait of Isabelle Simi that is surrounded by samples of her other lifelong passion—button collecting.

The Simi Winery produces 150,000 cases of vintage-dated varietal table wines yearly. It is particularly known for its Chardonnay and Cabernet.

Open all year: daily 10-4:30. Guided tours at 11, 1, and 3, 45 minutes. Free tasting. Retail outlet: wine, gifts. Picnic grounds. Annual 10-kilometer Simi Run to benefit Sonoma County Wine Library, last weekend in August. Access for the handicapped. Directions: Highway 101 to Dry Creek exit. At exit go east. Winery one mile north, on left.

Stephen Zellerbach Vineyard
4611 Thomas Road
Healdsburg, California 95448
Telephone: (707) 433-9463
Owners: Stephen and Cici Zellerbach

Modern Winery with 69-acre vineyard planted to Cabernet and Merlot near Field Stone Winery on Chalk Hill Road. Twenty-thousand cases annually of Cabernet, Merlot, and Chardonnay. Especially noted for Cabernet Sauvignon, '78 vintage in particular. Open all year: daily 10-5. Guided tours by appointment only (one week's notice suggested). Free tasting daily of all three wines produced. Retail outlet: wine and homemade grape jelly at harvest time. Picnic tables. November Open House (Thanksgiving weekend). Access for the handicapped. Directions: From Santa Rosa, take Highway 101 north to Shiloh Road exit. Turn left on Old Redwood Highway. Right on Pleasant Avenue. Left on Chalk Hill Road. 6½ miles on Chalk Hill to winery. Mailing address: 14350 Chalk Hill Road, Healdsburg, California 95448.

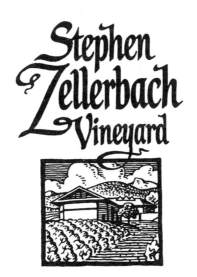

Kenwood

Chateau St. Jean Vineyards and Winery
8555 Sonoma Highway (P.O. Box 293)
Kenwood, California 95452
Telephone: (707) 833-4134
Owner: Suntory, International

In a manicured park-like setting, set against the Sugarloaf Ridge at the northern end of the Sonoma Valley, lies Chateau St. Jean , named for the wife of one of its three original founders. Originally the site of the Goff family country home, built in the Roaring Twenties, the property now contains, in addition to the refurbished mansion, a winery built in the same architectural style, complete with tower, arches, red-tile roof, and stuccoed exterior. A formal entrance road—flanked by seventy-seven acres of vineyards planted to Chardonnay, Pinot Blanc, Gewürztraminer, Johannisberg Riesling, and Sauvignon Blanc—leads to the winery. (These grapes represent only twenty percent of those needed for production. The difference is either purchased or grown on land leased off the site.)

The tour here is self-guided, with signs, photographs, and other visual aids posted along the way to give you a good idea of what's happening in the winery. The visitors' path through the working facility is via an interior bridge. At its end, stairs lead to the top of the tower where you can enjoy splendid views of the surrounding vineyards.

Tastings are offered in the Goff mansion, a few steps away. Usually four wines can be tasted daily, except when the winery is especially busy and only three are opened. Free samples of Chateau St. Jean sparkling wine are also sometimes available. (Come during the week, first thing in the morning, since only two bottles are opened—and when those are emptied, that's it!)

The winery boasts splendid grounds, perfect for a picnic lunch. There are ample picnic tables within a grove of shade trees, and park benches within a profusion of flowers face the fountain in the square.

Chateau St. Jean's production is devoted almost exclusively to 100 percent varietal vintage-dated and vineyard-designated white table wines, both still and sparkling. (Some vin blanc is produced, but this too is made from premium white varietal grapes.) The winery is particularly noted for its Chardonnay, Late Harvest Riesling, Gewürztraminer, Sauvignon Blanc, and sparkling wine, the last of which is produced by the French *methode champenoise.*

Open all year, daily except major holidays: 10-4:30. Self-guided tour, 30 minutes. Free tasting. Retail outlet: wine. Extensive picnic facilities. Access for the handicapped. Directions: From Sonoma, Highway 12 north. On east side of road, 10 miles from town center.

Sebastopol

Dehlinger Winery
6300 Guerneville Road
Sebastopol, California 95472
Telephone: (707) 823-2378
Owner: Tom Dehlinger

Small premium winery (7000 cases) and vineyard (31 acres) in the lower Russian River region of Sonoma County. Founded in 1975 by Tom Dehlinger. Dehlinger, a winemaker by profession, studied viticulture and oenology at UC Davis, apprenticing afterwards at Hanzell, Beringer, and Dry Creek wineries before starting his own firm. Production limited to varietal table wines made largely from estate-grown grapes. Wines produced are Chardonnay, Pinot Noir, Zinfandel, and Cabernet Sauvignon. Open all year: Monday-Friday, 1-4:30; Saturday and Sunday, 10-5. Guided tours by appointment only. Free tastings daily. Retail outlet: wine. Holiday Bazaar (the three days after Thanksgiving). Directions: West at Highway 12 exit from Highway 101 (at Santa Rosa). Follow Highway 12 west, 8 miles, to Sebastopol. At Sebastopol, 2 miles north on Guerneville Road. Winery is located on northeast corner of Guerneville and Vine Hill Roads.

Iron Horse Vineyards
9786 Ross Station Road
Sebastopol, California 95472
Telephone: (707) 887-1507
Owners: Barry and Audrey Sterling, Forrest and Kate Tancer

Fully self-contained wine estate, "on top of the world," north of Sebastopol in the sharply rolling countryside of the Russian River Valley, Sonoma County. Named for miniature railroad that once traversed the property. One hundred and forty acres planted to Chardonnay, Cabernet, and Pinot Noir. Centerpiece is the restored Victorian Sterling domicile and gardens. Tours of the winery and tasting, both by appointment only. These cover the whole operation, including riddling, disgorging, and dosage of the sparkling wines.

The walk down the hill, along the edge of the vineyards, towards the winery buildings which were built in the 1970s and now reflect subsequent add-ons, offers panoramic views and a sense of the living aspect of wine making—the grapes themselves. Production: 24,000 cases annually; fifty percent sparkling wine, fifty percent still table wine. Especially known for '81 sparkling wines: Blanc de Noir and Blanc de Blanc, and '81 still table wines: Cabernet and Pinot Noir. Open all year: 9-4; Monday-Saturday, May-October; closed Saturdays November-April. Guided tours and tasting by appointment only. Retail outlet: wine. Most areas accessible for the handicapped. Directions: From Santa Rosa, north on Highway 101 to Guerneville Road exit. West on Guerneville Road to Highway 116. North on 116 approximately one mile to Ross Station Road. Left on Ross Station to winery (at end of road).

Sonoma

Buena Vista Winery
27000 Ramal Road (P.O. Box 182)
Sonoma, California 95476
Telephone: (707) 938-8504
Owner: A. Racke

For anyone interested in retracing the roots of California's wine industry, a pilgrimage to the historic Buena Vista Winery, just outside downtown Sonoma, is a must. The attractive property incorporates the 1862 Press House and 1857 cellars built by Count Agoston Haraszathy, the acknowledged father of California's premium wine industry. (The Count was dispatched in 1861 by the Governor of California to the important wine-growing countries of Europe to bring back cuttings of vinifera vines. He imported 100,000 vines, embracing almost 300 different varieties, which were distributed throughout California.)

The original stone cellars and Press House, registered California historic landmarks, now serve as a visitors' center for the Buena Vista Winery. (The functional winery and vineyards are located on a 620-acre estate ten miles away in the Carneros region.) On the ground floor of the Press House is the main tasting room and retail sales area. Upstairs, a balcony provides gallery space for changing exhibitions of paintings and crafts. The tour here begins in the cellars, where photographs show the winery as it was in the Count's days and illustrate modern harvesting and production methods. Beyond the educational exhibit, the wine caves hewn out of the hillside by Chinese laborers are visible, and one ends the circle at yet another tasting room, housing the champagne bar and vintage wines library. The rare and reserve wines on display in the library are for sale, and a free tasting of the French champagne named for Haraszathy's son, Arpad, and made by Buena Vista's parent company, is offered.

Outside the handsome stone buildings are extensive picnic facilities. During the summer months, the courtyard is the setting for a round of cultural events including all-Mozart concerts and Shakespearean plays.

Buena Vista wines have been served at White House presidential state dinners. Of the firm's varietal table wines, the Cabernet Sauvignon is considered its flagship wine. Buena Vista also produces two proprietary wines: Spiceling which is sixty percent Gewürztraminer, forty percent Riesling; and Pinot Jolie, Early Harvest Pinot Noir. And when there is a surplus of wine grapes, Buena Vista also frequently bottles the unfermented juice, which is delicious.

Open all year, daily except Thanksgiving, Christmas, New Year's Day: 10-5. Self-guided tours. Free tasting. Retail outlet: wine, gifts. Art gallery. Special events: concerts, plays, festivals. Picnic facilities. Access for the handicapped. Directions: From Sonoma town square, take East Napa Street over railroad tracks. At East Eighth Street take a left. At Old Winery Road follow signs to Buena Vista.

Gundlach-Bundschu Winery
2000 Denmark Street
Sonoma, California 95476
Telephone: (707) 938-5277
Owner: Jim Bundschu

Historic family-owned stone winery founded in 1858 by Jacob Gundlach. Located on 350-acre estate, the "Rhinefarm." Premium winery specializing in estate-grown-and-bottled varietals, especially Cabernet Sauvignon, Merlot, Chardonnay, and Gewürztraminer. Open all year, daily except major holidays: 11-4:30. No tours. Free tasting. Retail outlet: wine and gifts. Picnic area. Directions: From Sonoma town square, go east on Napa Street. Turn right on Eighth Street. East, then turn left on Denmark Street. Proceed on Denmark for approximately one mile, taking road through the vineyard as marked. Mailing address: P.O. Box 1, Vineburg, California 95487.

Hacienda Wine Cellars
1000 Vineyard Lane (P.O. Box 416)
Sonoma, California 95476
Telephone: (707) 938-3220
Owners: A.C. Cooley, F.H. Bartholomew, S.W. MacRostie

Hacienda Wine Cellars, like its neighbor Buena Vista, is also located on part of the original Haraszathy tract. Although much smaller than Buena Vista, the winery is housed in an attractive Spanish-Colonial brick building. Built in 1926 as a home for wayward women, the structure later served as the Sonoma County Hospital. Tours of the winery are by appointment only. But if you stop by unannounced, a glimpse of the barrel aging cellar is visible through a decorative grill in the tasting room.

The tasting room/retail sales area is decorated with old prints of the Haraszathy estate. Six wines are usually available for sampling daily. Outside is a charming "wine garden" with many picnic tables. These are situated on a knoll overlooking a duck pond and the historic Haraszathy vineyards.

Hacienda produces vintage-dated varietal table wines almost exclusively. (Fortified wine accounts for approximately one percent of its output.) The winery is particularly noted for its Gewürztraminer, Sauvignon Blanc, and Cabernet Sauvignon.

Open all year, daily except New Year's Day, Easter, Thanksgiving, and Christmas: 10-5. Guided tours, by appointment only. Free tasting. Retail outlet: wine and gifts. Picnic facilities. No access for the handicapped. Directions: From Sonoma town square, go east on East Napa Street. Turn left at East Seventh and follow signs to winery.

Sebastiani Vineyards
389 Fourth Street East (P.O. Box AA)
Sonoma, California 95476
Telephone: (707) 938-5532
Owners: The Sebastiani Family

Due east of Sonoma's peaceful town square—ringed with specialty shops, restaurants, and palm trees—are a clutch of interesting and historic wineries, the largest and closest of which (within walking distance) is the Sebastiani Vineyards.

Sebastiani was founded in the early 1900s by Samuele Sebastiani who bought the property (then the Milani Winery) with money he earned hauling cobblestones from Sonoma to pave the streets of San Francisco. His son, August, succeeded him, and now Sam, his grandson, is president. While the company has continued in family hands for three generations, its style of winemaking and its product line have kept pace with the times and current tastes. Sebastiani has grown from a manufacturer of bulk wine sold under other labels during Samuele's ownership, to a producer of prizewinning vintage-dated varietals—the Sam Sebastiani label—under its current management. Its yearly output is an astonishing 2.2 million cases.

The half-hour tour of the winery begins in the circa 1913 stone livery stable, now an aging cellar, where some of the original equipment from Samuele's era is still on view. Modern winemaking facilities on view include the storage area, crushing area, bottling line, and more, but the most memorable aspect of the visit is the winery's unparalleled collection of carved oak and redwood casks.

The fine carvings, which are not solely relegated to casks alone, but adorn the walls, mantels, and almost any available wood trim, are the work of Earle Brown, a signmaker, who spent his retirement years, 1968-1984, creating them. The carvings, which have wine themes, are instructive as well as decorative. One sequence depicts the different grape varieties with the leaves of the grape clusters exhibiting as much detail as a pen and ink drawing. Another series, the "Vintner's Calendar," illustrates the cycle of grape to wine during the year.

The tour of the winery ends in the tasting room, where Sebastiani's many medals and ribbons are displayed. Nearly twenty different wines are available each day for sampling. Be sure to walk down to the end of the bar where you can actually step into half an enormous oak tank. Then look up; above is a display of antique barrel-making tools. Not part of the tour, but open to the public in an adjoining room, is the Rose Gaffney collection of Indian artifacts, including beads, arrowheads, and baskets of several California and Southwestern tribes.

A UNIQUE   VARIETAL ROSÉ
CELEBRATING THE   NATURAL PINKISH HUE
OF THE GEWÜRZ   TRAMINER GRAPE.

ROSA

VINEYARDS ESTABLISHED 1825

Sebastiani

1983

SONOMA VALLEY

GEWÜRZTRAMINER

PRODUCED AND BOTTLED BY SEBASTIANI VINEYARDS
SONOMA, CALIFORNIA ALC. 12.9% BY VOL.
BONDED WINERY 876

Sebastiani has a long product line, but basically the Sebastiani Vineyards label appears on premium wines and proprietor's reserve wines bottled in the standard French bottle size (750 milliliters), while the August Sebastiani label appears on generic and varietal wines packaged in magnum bottles (1.5 liters, or twice the standard size) and even larger sizes.

Open all year, daily except major holidays: 10-5. Guided tours, daily from 10-4:20, 30 minutes. Free tasting. Retail outlet: wine, gifts, books. Picnic tables. Indian artifacts museum. Nouveau Gamay Beaujolais ceremony, November 15 every year. Access for the handicapped. Directions: From Sonoma town square, east on East Spain Street three blocks to Fourth Street East. (There are signs from the town plaza.)

Windsor

Landmark Vineyards
9150 Los Amigos Road
Windsor, California 95492
Telephone: (707) 838-9466
Owners: The Mabry Family

Small family-owned premium winery, distinguished by its formal entrance, an avenue lined with stately 100-year-old cypress trees. The trees are a local landmark, whence the winery's name. Tasting room and retail sales housed in a graceful Spanish-style structure, formerly a residence. Noted for its Chardonnay. Guided tours, tastings, and sales by appointment only (one day's notice suggested): Monday-Friday, 9-5. "Sundays at Landmark": a series of plays, concerts, and exhibitions held monthly during spring and summer. Picnicking in the gardens. Directions: North from Santa Rosa on Highway 101 to first Windsor exit. Winery to the east of 101 on frontage road.

Sonoma-Cutrer Vineyards
4401 Slusser Road
Windsor, California 95492
Telephone: (707) 528-1181
President: Brice Cutrer Jones

The turn off River Road to Slusser, which ultimately leads to Sonoma-Cutrer Vineyards, is not well marked, nor may it ever be. For the owners of this spanking new ultra-modern winery (they were still spackling the sheetrock and laying the sod when I visited) are not particularly interested in turning their stunning facility into a major tourist attraction. But if you are seriously interested in seeing what a "no-compromise, state-of-the-art facility" looks like, make an appointment. This is the stop for you!

Designed by architect John Miller, the multileveled winery is built into a crest of land alongside a pond stocked with bass, exotic ducks, and geese. The exterior of the winery is Sonoma-red wood siding, a color and material chosen to be reminiscent of the hop kilns and barns which once dotted the countryside. A massive Prussian-blue trellis symbolizes the agricultural shift from hops, cattle, and farming in the area to vineyards and winemaking. And the world-class croquet courts located below the front terraces (and the site of the 1985 Western Regional United States

Croquet Association Tournament) represent a personal interest of the winery's president, Brice Jones.

Sonoma-Cutrer was built to produce only Chardonnay—a Chardonnay that will continue to improve in the bottle for ten years. To this end the facility has incorporated into its design some innovative approaches to winemaking. It boasts the only grape cooling tunnel in the world and a natural "breathing" earth floor in its aging cellar, a first in the United States.

The visitor's tour of Sonoma-Cutrer will take in the above, as well as the galleria, a skylight-illuminated corridor, where an educational exhibit and Andy Warhol drawings of grapes will be hung. From the galleria, the bottling line, sorting tables, and other winemaking facilities will be seen. At the conclusion of the tour, tastings will be enjoyed in the visitors' reception area with its massive polished-granite fireplace.

Chardonnay is Sonoma-Cutrer's only wine, but a limited quantity of *methode champenoise* champagne/sparkling wine (they haven't decided what to call it yet), and also made from the Chardonnay grape planted on their surrounding 700 acres, is planned for the future.

Open all year. Tour, tasting, and sales by appointment only. Tour, and tasting which follows, takes approximately one hour (one week's notice preferred). Retail outlet: wine. Access for the handicapped. Directions: Highway 101 north from Santa Rosa to River Road exit. West on River Road for 3½ miles to Slusser Road. Right on Slusser Road to winery on your left.

MENDOCINO AND LAKE COUNTIES

Hopland

Fetzer Vineyards
13500 South Highway 101
Hopland, California 95449
Telephone: (707) 744-1737
Owners: The Fetzer Family

Tasting room located in recycled Hopland High School (address above), combines a deli/sandwich shop, gift shop, and gallery with wine tasting. Adjoin-

ing gourmet restaurant under separate management. Vineyard and winery, 30 miles north, in far reaches of Mendocino County. Family owned and operated: ten of the eleven Fetzer children and their mother, Kathleen, manage the winery founded by the late Bernard Fetzer in 1968. A 600,000-case winery known for its sturdy red wines, especially vineyard-designated Zinfandel. Open all year, except Christmas: 9-5. Free tastings. Retail outlet: wine, gifts, picnic fare. Deli and gourmet restaurant. Picnic and barbecue facilities. Harvest Festival. Access for the handicapped. (Guided tours of winery and vineyard by appointment only, spring preferred [one week's notice], call winery: (707) 485-7634.) Directions to tasting room: on Highway 101 in downtown Hopland. Mailing address: P.O. Box 227, Redwood Valley, California 95470.

McDowell Valley Vineyards
3811 Highway 175 (P.O. Box 449)
Hopland, California 95449
Telephone: (707) 744-1053
Owners: Richard and Karen Keehn

Oenophiles of an environmental bent will want to take in McDowell Valley. Like Fetzer, it has opened a tasting room in downtown Hopland. But if you have the time, take the four-mile drive east on narrow, winding Highway 175 to its home winery and vineyards. The short detour will be well worth your while, for McDowell Valley Vineyards is one of very few solar-powered wineries in the world.

The modern two-story wood and concrete winery was built by the Keehn family in 1979 on the floor of McDowell Valley, a small narrow valley just east of Hopland. The 600-acre estate, originally settled by Paxton McDowell in 1852, was largely planted by the early 1900s, and in the 1970s the land was purchased by Richard and Karen Keehn, native northern Californians with eight children.

The small 65,000-case estate winery is a family operation. Its picturesque hilltop location is graced by a bowl-shaped lawn for picnicking and concerts, and, just beyond, 360 acres are planted to vineyards, primarily Sauvignon Blanc and the true French Syrah.

Visitors to the winery are struck by its unique architectural feature: the rooftop solar collectors, which face south, are immediately visible. These provide enough energy from the sun's rays for the heating, cooling, sterilization, and hot water needs of the winery. Tours of the working facility are by appointment only, but visits to the grounds and tasting room alone do not require advance notice. The second-story tasting room and decks afford not only a spectacular view of the seemingly endless sea of vineyards, framed by mountains and two lakes in the distance, but some glimpses into the winery below as well.

McDowell Valley wines are all estate grown and bottled. The focus is on vintage-dated varietal table wines, with the true Syrah the firm's most outstanding product. (It won four gold medals in 1983.) But McDowell Valley also produces cheaper proprietary wines, both red and white. The latter are also vintage dated.

Usually ten to twelve wines are available for tasting daily. You can enjoy the McDowell product either in the tasting room or with lunch, if you happen to pack one, on the adjoining decks furnished with redwood picnic tables.

Open all year: daily 10-5. Guided tours by appointment only. Free tasting. Retail outlet: wine, gifts. Picnic deck and lawn. Special events: Harvest festival, Sept.; Dinner with the Winemaker. Directions: From Hopland, go east on Highway 175 for 4 miles.

Middletown

Guenoc Winery
21000 Butts Canyon Road (P.O. Box 1146)
Middletown, California 95461
Telephone: (707) 987-2385
Proprietor: Orville Magoon

"Join me in Paradise!" were the words Lillie Langtry used to describe her newly acquired property, the Guenoc Ranch in Lake County, to General W. H. L. Barnes, the attorney responsible for purchasing it for her in 1888. The property belonged to the celebrated English beauty and actress for eighteen years. And the lovely country Victorian house, together with its Langtry associations and memorabilia, is the focus of the modern winery that makes its home here nearly a century later.

Mrs. Langtry was a horse lover. She and her paramour, Freddie Gebbhard, who owned a neighboring ranch, stocked their properties with purebreds and set up a racetrack. But Lillie was also interested in winemaking. To this end she imported a French winemaker, planted a small vineyard, and produced, according to agricultural records dated 1891, "fifty tons of Burgundy wine." With Prohibition her foray into winemaking ended.

Purchased by the Magoon family in the early '60s, Lillie's Victorian country home, perched on a hill, has been fully restored. The graceful second-story balcony, which once wrapped around the exterior of the house but was removed by subsequent owners, has been fortuitously replaced. From it you can stand and look out over the Guenoc Valley and Mrs. Langtry's vineyard, now replanted and expanded to an additional 250 acres by the current owners. These are planted to Chardonnay, Chenin Blanc, Sauvignon Blanc, Semillon, Zinfandel, Petite Sirah, Cabernet Sauvignon, Petite Verdot, Cabernet Franc, Malbec, and Merlot.

The interior of the house has been adapted to include a wine tasting room—in Lillie's original dining room—an office for visitors, and guest accommodations.

Visitors to the winery are met by the winery staff at the parking lot, where the tour begins. On view are the modern winery, the vineyard, and the tasting and sales area where an exhibit of Lillie Langtry memorabilia is displayed.

Guenoc's product line, 50,000 cases annually, is devoted solely to red and white still table wines. The winery is particularly noted for its Zinfandel, Chardonnay, Sauvignon Blanc, Chenin Blanc, and Petite Sirah.

Open all year: Thursday-Sunday, 10-4:30. Guided tours: 10-4:30. (Telephone ahead in winter months since roads may be closed because of heavy snows.) Tasting. Retail sales: wine, gifts. Collection of Lillie Langtry memorabilia displayed. Picnic facilities overlooking the Detert Reservoir and Guenoc Valley. Tasting room accessible to the handicapped. Directions: From Middletown, take highway 29 north 6 miles to Butts Canyon Road. Right turn on Butts Canyon to the Guenoc Winery sign (on your left). Detert Reservoir, a very large lake, will be on your right. Turn left and drive up hill through the winery gates.

Philo

Edmeades Vineyards
5500 California State Highway 128 (P.O. Box 177)
Philo, California 95466
Telephone: (707) 895-3232
Owner: Edmeades, Inc.

One of the first vineyards planted in the Anderson Valley after Prohibition, Edmeades, located north of the small town of Philo on a former apple ranch, was founded by Donald Edmeades, a Pasadena physician, who began planting the thirty-five acre vineyard in 1964. The winery itself was founded in 1972 by Edmeades' son, Deron, in a converted apple dryer on the property. A rustic tasting room is featured. Best known for Cabernet Sauvignon and for producing California's first, and only, ice wine, a French Colombard, 1977—now a collector's item. Known, too, for proprietary wines, including Rain Wine (whose name reflects the fact that Anderson Valley is the rainiest viticultural region in California) and Whale Wine. A percentage of the income from the latter is donated to marine research and preservation. Open all year: Winter, 11-5; Summer, 10-6. Guided tours by appointment only (48 hours notice required), 30 minutes. Free tasting. Retail outlet: wine, gifts. Picnicking. Summer concerts in the vineyards feature jazz, rock, and country music. Access for the handicapped. Directions: Three miles north of Philo on west side of Highway 128.

Husch Vineyards
4900 Star Route
Philo, California 95466
Telephone: (707) 895-3216
Owners: The H.A. Oswald Family

A small family-owned operation devoted to the production of premium estate-bottled varietal wines (12,000 cases annually). Winery is the oldest in the Anderson Valley. Its cellar was bonded in 1971, by Gretchen and Tony Husch, only one year earlier than its neighbor, Edmeades. Eight years later the vineyards and winery were bought by UC Davis-trained winemaker and Mendocino Conty grape-grower of over twenty years, Hugo Oswald III. Since then the winery has been run by Hugo with the help of six family members. Husch vineyards is known for its white wines, particularly Sauvignon Blanc, Chardonnay, and Gewürztraminer. Visitors to the winery are invited to chat leisurely with the staff, stroll through the vineyards, or picnic at a table under the grape arbors. The pleasant ambiance is best described as "unadorned rustic." Open daily all year: Winter, 10-5; Summer, 10-6. Tours: guided, time permitting, or self-guided (reservations suggested). Free tasting. Retail outlet: wine. Picnic tables. Tasting room accessible to the handicapped. Directions: Five miles north of Philo on west side of Highway 128.

Navarro Vineyards And Winery
5601 Highway 128 (P.O. Box 47)
Philo, California 95466
Telephone: (707) 895-3638
Owners: Ted Bennett and Debra Cahn

Navarro Vineyards and Winery looks as if it should be in Bavaria rather than in the heart of the Anderson Valley. This charming chalet-style winery, surrounded by flower gardens and grape vines, was built entirely from lumber milled from three giant redwoods felled on the property. A small 12,000-case winery, founded in 1974 by Ted Bennett and Debra Cahn, Navarro prefers to stay just the size it is so it can continue to produce a high-quality product from the finest grapes.

About sixty-five percent of the grapes needed for Navarro's production are grown in the fifty acres of vineyards surrounding the winery; the rest are purchased from local growers. Of the estate acreage, twenty-seven are planted to Gewürztraminer, a prime interest of Ted Bennett's; fifteen to Chardonnay; and about eight to Pinot Noir.

Tours are by appointment only, and only when there is staff free to accommodate visitors. But a good selection of wines is available for tasting daily—usually five. And occasionally two wines made from the same grape variety are offered so you can compare the differences in how they were made. Glass doors from the tasting room lead to a picnic deck which extends into the vineyards. Furnished with tables and chairs (shaded by colorful umbrellas), the deck is a lovely place to stop for lunch.

Navarro is known for its varietal dinner wines, and especially its Germanically styled Gewürztraminer. Proprietary wines include a vin blanc, vin rouge, and Edelzwicker. The winery also produces a very popular non-alcoholic grape juice made from spicy Johannisberg Reisling grapes.

Open all year, daily except Christmas and Thanksgiving: 10-5, Winter; 10-6, Summer. Guided tours by appointment only, and only when there is staff available (write or call as far ahead as possible), 15-30 minutes. Free tastings. Retail outlet: wine, wineglasses. Picnic deck. Access to tasting room for the handicapped. Directions: Three miles north of Philo on east side of Highway 128.

Ukiah

Parducci Wine Cellars
501 Parducci Road
Ukiah, California 95482
Telephone (707) 462-3828
Management: The Parducci Family

Parducci Wine Cellars, just a little north of Ukiah on Highway 101, has been called the patriarch of Mendocino County wineries, a title that suits it well. The oldest winery in the county, Parducci was founded more than fifty years ago by Adolph Parducci on a 100-acre tract of land now called the "Home Ranch."

Under Adolph's leadership, the Parducci label was associated primarily with bulk wines. But, in 1964, when his sons, George and John, took over the family business, the winery's focus was changed to premium wines, and many of its substantial vineyard holdings were replanted to these grape varieties. Today, although a majority of its stock is owned by Teachers Management Institute, Parducci is still a family-run operation, with John continuing as winemaker and George as manager.

Visitors to Parducci will tour a 250,000-case winery housed in a cluster of buildings, the newest of which contains the bottling line and case storage facilities. Just beyond, a group of older structures house the rest of the winery operations, including the stainless-steel fermenters and old redwood and oak cooperage. The guided tour ends in the large tasting/gift/retail sales area and gallery located in a building redolent of Spanish architecture. After sampling, lunch can be enjoyed on an adjoining outdoor picnic patio.

In addition to the "Home Ranch," Parducci Wine Cellars owns vineyards in Talmadge and Largo, south of the winery, that are planted to Sauvignon Blanc, Chenin Blanc, Riesling, French Colombard, Chardonnay, Cabernet Sauvignon, Petit Sirah, Pinot Noir, and Zinfandel.

Parducci is noted for its dry varietal table wines. Special lots are labeled "Cellar Master's Selections."

Open all year, daily except major holidays: 9-6; Winter hours, 9-5. Guided tours, daily: 9-4, 30 minutes. Free tasting. Retail outlet: wine, gifts. Art gallery. Picnic tables. Tasting room accessible to the handicapped. Directions: Take Highway 101 north from Ukiah to Lake Mendocino Drive exit. Exit right, go under highway, and follow signs to the tasting room.

BAY AREA AND NORTH CENTRAL COAST

Gonzales

Monterey Vineyard
800 South Alta Street (P.O. Box 780)
Gonzales, California 93926
Telephone: (408) 675-2481
Owner: The Seagram Classics Wine Company

The name "Monterey Vineyard" is something of a misnomer, for the winery does not own any vineyards but instead purchases from local growers all of the grapes required for its 140,000 to 160,000-case annual production. There is no question, however, about Monterey Vineyard's preeminence as a winery. Built in 1974, it was the first in Monterey County to offer daily tastings and tours of its cellars, and it is still one of the most instructive and attractive wineries to visit in the area.

A combination of old and new elements, the winery is located outside the village of Gonzales on portions of land originally granted in 1836 to Theodoro Gonzales by the Government of Mexico. The exterior of the winery reflects the Spanish and Mexican heritage of early California by combining in its architectural elements earth-toned stucco, red tile roofs, weathered ironwork, stained-glass windows, arches, and even a tower.

Inside, the winery is modern to the minute. And while the guided tours, conducted hourly from 11 to 4, reveal the usual array of crushers, presses,

fermenters, and oak cooperage, your guide will point out some of the innovative operational details added by Monterey's winemaker/president, Dr. Richard Peterson, informative details that inject a special dimension to the tour of the working facility. Tours end typically in the tasting room where the Monterey Vineyard product can be sampled.

The winery's production is devoted almost exclusively to table wine, but a small amount of Brut *methode champenoise* champagne is also made. The premium vintage varietal wines include Chardonnay, Fumé Blanc, Pinot Blanc, Chenin Blanc, Johannisberg Riesling, and Gewürztraminer. In addition, Monterey produces vintage varietal blends, namely Classic Dry White, Classic Red, and Classic Rosé. All of these, and Monterey's Brut champagne and Botrytis sweet dessert wines, can be purchased on the premises. The latter specialty wines, incidentally, are available *only* at the winery.

Don't miss the climb to the top of the observation tower and deck during your visit. It affords sweeping views of the scenic valley and the Gabilan and Santa Lucia mountain ranges in the distance.

Open all year: daily 10-5. Guided tours, on the hour, 11-4. Tasting. Retail outlet: wine, gifts. Picnic facilities next to a lake. Festivals (check with winery for dates). Access for the handicapped. Directions: Take Highway 101 south from Salinas. Exit at Alta Street. Take Alta, north, ½ mile to winery. (Winery is 17 miles south of Salinas on southern edge of village of Gonzales.)

Greenfield

Jekel Vineyard
40155 Walnut Avenue (P.O. Box 336)
Greenfield, California 93927
Telephone: (408) 674-5522
Owners: William and August Jekel

Jekel is a very modern winery clothed in a deceptively rustic barn facade. It was founded by identical twins Bill and Gus Jekel, on a 140-acre estate in Monterey County. This is the driest viticultural area in California, with an annual rainfall of only 9 to 10 inches. But the cooling Monterey Bay winds, abundant ground water for irrigation, and well-drained soil are compensating factors which contribute to slow maturation of the grapes (some harvested as late as Thanksgiving); the result is a wine of intense fruit with a balance of tannins.

The Jekels began planting their acreage to vineyards in 1972, concentrating on five varieties: Johannisberg Riesling, Pinot Noir, Pinot Blanc, Cabernet Sauvignon, and Chardonnay. Six years later they built the winery, a red structure whose architecture reflects the turn-of-the-century barns that dot the Salinas Valley. Since the winery was completed, it has been expanded twice to a present annual capacity of 50,000 cases.

Jekel Vineyard produces only vintage-dated, varietal table wines which are estate bottled. The vineyard is best known for its Riesling and Chardonnay, but its Cabernet Sauvignon is also popular.

Open all year: Thursday-Monday, 10-5. Guided tours at 10 and 5. Free tasting. Retail outlet: wine, gifts. Picnic facilities. Access for the handicapped. Directions: From Salinas, follow Highway 101 south 32 miles to Walnut Avenue exit. West on Walnut Avenue one mile to winery (between 12th and 13th Streets).

Livermore

Concannon Vineyard
4590 Tesla Road
Livermore, California 94550
Telephone: (415) 447-3760
Owner: Distillers Company, Ltd.

Traveling south of the sleepy town of Livermore on North Livermore Avenue will bring you to the gates of two important Livermore Valley wineries—Concannon Vineyard and the neighboring Wente Bros. Both are rich in history and family tradition—their original buildings have both been designated California landmarks—and both are instructive and hospitable wineries to visit.

The first, appearing on your left through impressive wrought-iron gates and a clipped privet-lined drive, is Concannon Vineyard. Founded by James Concannon in 1883, it is a 90,000-case estate winery, surrounded by 180 acres of vineyards. Although the winery was bought in 1983 by Distillers Company, Ltd., of Edinburgh, Scotland, Jim Concannon, grandson of the founder, is still very involved in its management.

Eleven employees are trained to give the regularly scheduled twenty-five-minute tours of the Concannon facilities. These begin in the vineyards, where pruning techniques and other viticultural practices are demonstrated. (If your visit coincides with the right season, you will be permitted to taste the grapes.) The tour then progresses to the winery itself,

where the various winemaking operations, the aging cellar with its French and American oak barrels and stainless-steel tanks, and the bottling line are viewed. The tour ends in the windowless tasting room (it's due to be remodeled soon) for a sampling of the complete line of Concannon wines.

All of Concannon's production is table wine, two-thirds of which are varietal wines and the remaining third devoted to generics. Grapes are both estate grown and purchased. The winery is particularly known for its Cabernet Sauvignon, Sauvignon Blanc, Chardonnay, and Petite Sirah.

At the edge of the Concannon vineyards, and with views of the Livermore Mountains on the horizon, are picnic facilities for 200 to 300 visitors.

Open all year, except New Year's Day, Easter, Thanksgiving, and Christmas; Monday-Saturday: 9-4:30, Sunday: 12-4:30. Guided tours: June-August, Monday-Friday at 11, 1, 2, and 3; Saturday and Sunday at 12, 1, 2, and 3. September-May tours by appointment only. Free tasting, with or without tour. Retail outlet: wine, gifts. Extensive picnic facilities. Access for the handicapped. Directions: Forty miles southeast of San Francisco. Take North Livermore Avenue exit off Interstate 580 East. Proceed south 3 miles (there are signs to winery) through town on South Livermore Avenue to winery gates. (Tesla Road is an extension of Livermore Avenue.)

Wente Bros.
5565 Tesla Road
Livermore, California 94550
Telephone: (415) 447-3603
Owners: The Wente Family

A mile or so down the road from Concannon Vineyard is Wente Bros. winery, family owned and operated since its founding in 1883 by Carl H. Wente, a German immigrant. Wente Bros. is among California's biggest wineries, with a 600,000-case annual output and more than 1,800 acres of vineyard holdings in Alameda and Monterey Counties.

The tour of the Wente facility is one of the more extensive in the state—lasting forty-five minutes—and it is especially colorful in the fall during harvest, when you can taste the ripe grapes and watch the flurry of activity surrounding the crush. Tours (offered on weekdays only) are guided by informed personnel and commence at the California Historic Landmark plaque in front of the winery's new tasting room. The large working winery, with its storage tanks and bottling line, is open to visitors, as are the crushing and fermentation facilities. Some of the vineyards are located nearby, permitting close inspection.

After the tour, the full line of Wente wines may be tried in the spacious tasting room. The wines are poured by an unusually knowledgeable staff of employees, three of whom have been with Wente Bros. for several decades. In addition to the wines, an excellent selection of wine-oriented books is offered for sale. Interesting exhibits, including a corkscrew collection and vintage photographs documenting four generations of the Wentes' viticultural skills, are displayed along the walls.

Wente's best sellers are its Grey Riesling and Le Blanc de Blancs. One of the new blush wines, Gamay Beaujolais, is also popular. In 1983, to commemorate its centennial, Wente introduced its first sparkling wine, a *methode champenoise* brut. A second Wente winery, Wente Bros. Sparkling Wine Cellars, is scheduled to open soon to produce this wine alone. Located in the historic, newly renovated Cresta Blanca Winery buildings on nearby Arroyo Road, the property will contain extensive visitor facilities. A combination museum and tasting room and a gourmet restaurant with outdoor dining patio are to be surrounded by elaborate landscaping, incorporating picnic facilities, winery artifacts, and sweeping lawns. In addition, comprehensive tours of the disgorgement and tirage buildings will feature visits to the newly strengthened historic tunnels.

Open all year. Guided tours: Monday-Friday at 9:30, 10:30, 11:30, 1, 2, and 3. Free tastings: Monday-Saturday 9-4:30, and Sunday 11-4:30. Retail outlet:

wine, gifts. Patio for picnicking. Harvest Festival, three days over Labor Day Weekend. Access for the handicapped. Directions: Beyond Concannon on right, going east on Tesla Road.

Los Gatos

Novitiate Winery
300 College Avenue (P.O. Box 128)
Los Gatos, California 95030
Telephone (408) 354-6471
Owner: California Province of the Society of Jesus

Novitiate Winery is one of three church-owned wineries in the United States. (The others are northern California's Christian Brothers and Indiana's Saint Meinrad Archabbey and Theological Seminary). It was founded in 1888 by the Jesuits and is one of oldest continuing operating wineries in the United States.

The white partly Mission-style sandstone winery sits on a hill in a heavily wooded, mountainous area just southwest of the town of Los Gatos near Route 17. The winery is the smaller of the two structures seen from the road, the larger being the Sacred Heart Novitiate which was established in the late nineteenth century to train young men as priests and brothers in the Jesuit order. Novitiate Winery, founded shortly thereafter, was to provide sacramental wine for serving mass, and sales of its surplus product were to produce income to support the novices during their training. The arrangement apparently worked well; the Jesuit brothers became such excellent winemakers that they eventually turned to the production of dessert and then table wines for public consumption.

A visit to Novitiate will reveal a winery where little has changed since its founding years. Its ambiance can best be described as "authentic." Dim cellars hung with moss, cobwebs, and dust confirm your image of how century-old cellars should look. The cooperage is mostly wood, old, and handmade. (Very little stainless steel is evident.) The tour of the facilities, which takes about half an hour, is led by a Jesuit brother, clothed in the long robes of the order, who will lead you through the whole wine-making operation—from the arrival of the grapes at the winery to the final bottling process.

Tasting of Novitiate wines takes place in one of the old wine vaults. Its arched ceilings and thick moss-covered walls make it a cool and fitting ending to your visit. Six wines a day can usually be sampled.

In addition to wine for sacramental purposes, Novitiate's production

(50,000 cases annually) includes sweet dessert wines, red and white varietal dinner wines, generic wines, and sparkling wines. Specialties include Black Muscat, considered by connoisseurs one of the rarest of dessert wines; Black Rose table wine, made from the Muscat Hamburg grape; and Black Rose Champagne.

The winery no longer grows its own grapes, its once surrounding vineyards having succumbed to the suburban growth of Santa Clara County. Some immediate vines remain for atmosphere. Picnicking is permitted on the paved terrace.

Open all year, daily except major holidays: 10-4:30. Guided tours: Monday-Friday at 1:30 and 2:30, Saturday and Sunday at 11 and 1 (30 minutes). Tasting with or without tour. Retail outlet: wine, gifts. Picnic terrace. Access for the handicapped. Directions: From San Jose, take Highway 17 south to Los Gatos exit. Proceed east to Los Gatos Boulevard. Right turn on Los Gatos Boulevard (which becomes Main Street) to College Avenue. At College Avenue, go left, continuing up hill to winery property. (Winery is 10 miles south of San Jose.)

Mission San José

Weibel Champagne Vineyards
1250 Stanford Avenue (P.O. Box 3398)
Mission San José, California 94539
Telephone: (415) 656-2340
Owner: Fred Weibel

Weibel Champagne Vineyards has two locations: one in Ukiah in Mendocino County and the other here in the east Bay area of Livermore-Alameda. Of the two, this is the original facility, and by far the more interesting to visit, both for its rich historical associations and for its glamorous products—champagnes and sparkling wines.

The Weibel property here was originally the site of the historic Warm Springs Vineyard founded by Leland Stanford, railroad builder, California governor, and United States senator. In Stanford's time (the mid-1800s), a fashionable spa and hot springs flourished here. Both succumbed to earthquakes: the spa in 1868, the springs some years later.

Following the demise of the spa, Governor Stanford purchased the land surrounding the springs and planted 100 acres to grapes, founding a winery which he named Warm Springs. His brother, Josiah, successfully ran the business and bequeathed it to his son; the vineyards continued to prosper until the turn of the century, when phylloxera destroyed the vines. The property remained abandoned until 1945, when it was bought, nurtured, and expanded by Swiss champagne makers Rudolf Weibel and his son, Fred.

Your tour of Weibel will reflect both its historic past and the Weibels' modern additions. The entrance road, edged with eucalyptus and olive trees, leads to the quaint adobe tasting room. Informal tours of the Weibel winery begin here. Staff members cover the history of the winery at some length and give you a brief look at the warehouse, manufacturing, and bottling facilities located in nearby buildings.

Weibel is one of the few wineries that makes its champagne both in the traditional French, bottle-fermented *methode champenoise* and in the bulk *Charmant* process. After learning about the differences between the two processes, you can then see if your palate can discern the difference in the tasting room. (The complete Weibel product line, including the *methode champenoise* champagne, is usually available for sampling at no charge.)

Weibel has an extensive product line in addition to its champagnes and sparkling wines: varietal table wines, generics, dessert wines, and apéritifs are offered (the varietal table wines and generics are made at the Ukiah facility).

Open all year: 10-5. Guided tours: Monday-Friday only. Tasting. Retail outlet: wine, gifts. Special events: Sunday brunches, starlight dinners, and special tastings. Picnic facilities on patio with a view of the vineyards. Access for the handicapped. Directions: From Mission San José, take Highway 238 south one mile to Stanford Avenue. Left on Stanford Avenue one-half mile to winery.

San José

Almaden Vineyards
1530 Blossom Hill Road
San José, California 95118
Telephone: (408) 269-1312
Owner: National Distillers and Chemical Corp.

Almaden Vineyards, dating from 1852, vies with Buena Vista for the title of being the oldest continuously operating winery in California. Much larger than its Sonoma competitor—at 11 million cases a year it is the nation's sixth largest winery—Almaden's home winery is located in San José. Here, on Blossom Hill Road, the remaining estate vineyards, reduced now to only seventeen acres, buffer the winery from the suburban development at its door. The original adobe and brick winery building has been preserved. Situated in the midst of its more modern and more recent counterparts, it blends comfortably architecturally since all share similar Spanish profiles.

The approach to the Almaden holding does not reveal any of the winery operations. Instead, as you drive up to the gate and its guard house, you see formal French gardens designed by the noted landscape architect Thomas Church and set against a panoramic view of the Santa Cruz Mountains.

Tours of Almaden combine a history of its founder, Charles Lefranc, with the history of winemaking in California and the romance of wine. Along the route you see the remaining home vineyards, (Almaden has substantial vineyard holdings and winery operations in Monterey, Kern, Fresno, and San Benito Counties), the aging cellars with their oak and redwood cooperage dating to 1876, and its very impressive bottling facility, one of the largest and fastest in the industry.

Tours end in the original cellars which date back to the late 1850s. These adjoin the Louis Benoist Rose Garden. (Kay and Louis Benoist sold Almaden in 1967 to National Distillers.) Here a formalized tasting of Almaden wines is very cleverly conducted. Since the winery has an extremely long product line—over fifty different wines—the staff first defines your palate preference with a blind tasting of two wines, a sweet and a dry. Next, tasters are divided into two groups, according to style preference, where they are presented with three "win" wines of similar character. After these are tried, you get to sample white, red, rosé, and sparkling wines. Depending on how many people are in your group, allow 1½ hours for the combined tour and tasting.

Almaden's product line is very extensive and runs the full gamut from jug wine to fine vintage-dated varietals and sparkling wines. The firm is particularly known for its Cabernet Sauvignon, Chardonnay, Sauvignon Blanc, and dessert wines. One of its most popular wines is the pink Grenache Rosé. Its introduction in the '40s sparked the widespread interest in the American marketplace for rosé wines.

Open all year: March-October, 9:30-5:30; November-February, 8:30-4:30. Guided tours daily: March-October at 10, 11, 1, 2, 3, and 4; November-February at 10, 11, 1, 2, and 3. Tasting, with or without tour. Retail outlet: wine, gifts. Limited picnic facilities in Louis Benoist Gardens. Access for the handicapped. Directions: From San José, take Route 17 south to Camden Avenue. East on Camden Avenue, past Branham Lane, to Blossom Hill Road. Left on Blossom Hill to winery gates. (Don Pacheco Tasting Gardens at Almaden's Hollister facility, junction of Highways 152 and 156, open daily 10-5.)

Mirassou Vineyards
3000 Abron Road
San José, California 93135
Telephone: (408) 274-4000
Owners: Daniel, James, and Peter Mirassou

The current owners and operators of Mirassou Vineyards are the fifth generation to practice the wine-making tradition established here by their great-great-grandfather, Pierre Pellier. In the family tradition, the Mirassous are now handing their wine-making skills down to yet another generation.

M. Pellier came to California in 1854 in search of gold, but instead found his fortune in the eastern hills of the Santa Clara Valley. He bought land to plant vineyards and build a winery, importing cuttings of varietals from his native France to establish his crop.

Under the first Peter Mirassou, Pellier's grandson, the winery continued operating even during Prohibition, making ends meet, as did so many vineyards, by producing wine for medicinal and religious purposes. Upon Repeal in 1933, Peter's two sons, Edmund and Norbert, replanted the vineyards with premium varietal grapes such as Cabernet Sauvignon, Pinot Blanc, and Johannisberg Riesling. Only in 1966, after 112 years of winemaking, did the Mirassous begin marketing and distributing their wine under the family name.

Today, visitors will be particularly impressed by the innovative viticultural techniques and machinery recently implemented here, among them, the first permanent sprinkler-irrigation system to be used in vineyards, the first vineyard harvest-crusher, and most recently, the first mobile vineyard press. Tours of Mirassou's facility are given daily throughout the year, but are particularly interesting during the harvest season when the modern equipment can be seen in operation.

Mirassou produces 350,000 cases annually; 83 percent of its produc-

tion is still table wine, the rest sparkling wine. The winery owns 300 acres of vineyards at its San José location in Santa Clara County and an additional 1,000 acres in Monterey County. This acreage is planted to more than a dozen premium grape varieties including those for Mirassou's most popular wines: Chardonnay, Cabernet Sauvignon, Pinot Blanc, and Pinot Noir.

The Mirassou premises are the setting for myriad special events throughout the year. These include wine education classes, cooking classes, candlelight dinners, concerts, festivals, Sunday brunches, and even a grape run. The winery will provide a calendar of these activities upon request.

Open all year: Monday-Saturday, 10-5; Sunday, 12-4. Guided tours: Monday—Saturday at 10:30, 12, 2, and 3:30; Sunday at 12:30 and 2:30. Free tasting. Retail outlet: wine, gifts. Access for the handicapped. Directions: Highway 101 south from San José to Capitol Expressway exit. East on Capitol Expressway to Aborn Road (second light). Right on Aborn Road. Continue on Aborn approximately 2 miles to winery (on your right).

Saratoga

Congress Springs Vineyards
23600 Congress Springs Road
Saratoga, California 95070
Telephone: (408) 867-1409
Owners: Vic and Mabel Erickson, Dan and Robin Gehrs

In the heart of the lovely Santa Cruz Mountains is Congress Springs Vineyards, a small estate winery (5,000 cases) with an interesting history and a beautiful view of the surrounding countryside.

In 1892 a young Frenchman, Pierre Pourroy, cleared the land where Congress Springs now stands, planting grapevines where redwoods once towered, and, some thirty years later, completing a white masonry winery which he called Ville de Monmartre. M. Pourroy's twenty-five acres of vineyards produced a red claret (Zinfandel) and small quantities of a Riesling.

The winery closed during Prohibition and was not reopened until 1976 under the ownership of the Erickson and Gehrs families. The original Ville de Monmartre structure now serves two purposes. Its upper level is home to the winemaker Dan Gehrs, while the lower holds the winery and its storage facilities, including handmade oak cooperage dating from

1909 alongside new French and American white oak barrels. The expanded vineyards now cover fifty acres of the estate property and are planted to Chardonnay, Chenin Blanc, Pinot Blanc, Pinot Noir, Semillon, Gewürztraminer, Zinfandel, and Cabernet Franc. These estate-grown grapes account for seventy-five percent of the winery's production needs; the rest are purchased solely from local growers.

The tour of Congress Springs' small winery will take only about ten minutes. Either guided or self-guided, it covers the barrel storage room and fermenters. After you have toured and sampled the wine, have a snack at one of the winery's picnic tables, or spread a blanket on the expansive lawn and enjoy the beautiful view of Saratoga village three miles below. If you prefer more active pursuits, there are numerous walking trails and roads through the estate property to explore.

Congress Springs produces chateau-style premium wines. Included in its roster are the varietals Chardonnay, Pinot Blanc, Semillon, Chenin Blanc, Blanc de Noir, Zinfandel, Cabernet Sauvignon, and Pinot Noir; Chardonnay and Zinfandel are among the most popular.

The winery is also host to numerous special events throughout the year. These include a Mustard Festival in February; a Memorial Day Weekend Festival in May; and Meet-the-Winemakers events in August and September. It also publishes a quarterly newsletter, *The Grapeleaf,* available upon request at no charge. The Premier Tasting Club entitles its members to preview all new wines prior to their general release, offers special savings at the winery, and includes an annual summer picnic. Application forms are available.

Open all year: Friday, 1-5; Saturday and Sunday, 11-5. Tours, guided or self-guided, 10 minutes, no appointment. Free tasting all year; special tastings in mid-February and Memorial Day weekend. Retail outlet: wine. Picnic facilities. Walking trails. Special events. Access for the handicapped. Directions: 3½ miles above Saratoga village, west on Highway 9 (which is also called Congress Springs Road).

Soquel

Bargetto Winery
3535 North Main Street
Soquel, California 95073
Telephone: (408) 475-2258
Owners: The Bargetto Family

Bargetto Winery lays claim to being the largest (current output 35,000 cases annually) and the oldest continually operating winery in the Santa Cruz Mountains. It was founded in 1933 by brothers Philip and John Bargetto, who emigrated to the United States from the Piedmont region of northern Italy where their family had been growing grapes and making wine for more than 300 years.

The winery's early production reflected its owners Italian winemaking heritage. Under the direction of second- and third-generation Bargettos, Martin and David, who studied oenology and viticulture at UC Davis, more modern scientific techniques have been introduced. Reflecting these advances, the winery's product line now focuses on single vineyard-designated, vintage-dated premium varietal wines.

Bargetto grows no grapes, although it is currently looking for vineyard acreage. Instead, wine grapes are purchased from other vineyards, such as Tepusquet Vineyard in northern Santa Barbara County and the St. Regis Vineyard in southern Napa. Varietal table wines account for seventy percent of Bargetto's production; of these, its Chardonnay and Johannisberg Riesling are noteworthy. The remaining thirty percent of its output is devoted to some highly regarded natural fruit wines which are distributed nationally. These include raspberry, olallieberry, and apricot.

Bargetto's home winery and tasting room are located in a rustic barn-like structure erected on the banks of a picturesque creek. Informal guided tours of the complete winery facility take approximately half an hour. After the tour, visit the attractive tasting room, which is decorated with wine-making artifacts. Tastings of the Bargetto product can also be enjoyed in the adjoining outdoor courtyard, where the work of local artists is often exhibited.

Open all year: 10-5. Guided tours, weekdays at 11 and 2; weekends by appointment only. Free tasting. Retail outlet: wine, gifts. Picnic facilities in courtyard. Access for the handicapped. Directions: From Santa Cruz, take Highway 1 four miles south to Capitola-Soquel exit. At exit, go left under highway on Bay Avenue to Soquel Drive. Right on Soquel Drive, then left on North Main Street. Winery will be on your left. (Bargetto Winery also operates a tasting room/retail outlet/gift shop in Monterey. Called Bargetto's Cannery Row Tasting Room, it is located at 700 Cannery Row. Open daily at 10 for tasting and sales. Telephone: [408] 373-4053.)

SOUTH CENTRAL COAST AND SOUTH COAST

Creston

Creston Manor Vineyards and Winery
17 Mile Post, Highway 58
Creston, California 93432
Telephone: (805) 238-7398
Owners: David and Christina Crawford Koontz; Larry and
Stephanie Rosenbloom

Small (7,500 cases) premium winery in scenic La Panza Mountains of San Luis Obispo County. You're likely to see Christina Crawford, Joan Crawford's daughter, and her husband, David Koontz, in the vineyard or tasting room since they are partners in the winery. Short guided tours cover the grape receiving area, crusher, winery interior, fermentation tanks, barrel cellar and more. Ninety-five acres recently planted to Chardonnay, Sauvignon Blanc, Cabernet Sauvignon, and Pinot Noir. Product line: still table wine, vintage dated varietals only. Noted for its Sauvignon Blanc and Chardonnay. Picnicking, swimming, and fishing permitted at three lakes on the property. Open all year: daily 10-5. Guided Tours: 10-4, Monday-Friday (reservations recommended, one day's notice). Tasting. Retail outlet: wine, gifts. Directions: From Highway 101, 17 miles east on Highway 58.

Los Olivos

Firestone Vineyard
Zaca Station Road (P.O. Box 244)
Los Olivos, California 93441
Telephone: (805) 688-3940
Owners: Leonard K. Firestone, A. Brooks Firestone, Suntory
International

Firestone Vineyard, with an annual production of 75,000 cases, is the largest winery in the Santa Ynez Valley. The winery was founded in 1973 on vineyard acreage planted by former U.S. Ambassador to Belgium Leonard K. Firestone. The winery and vineyard farming operation is now managed by his son, Brooks, and is owned in equal partnership by the

two Firestones and Suntory International, Japanese whiskey distillers.

Firestone wines are all 100 percent varietal, vintage dated, and estate grown. Some are vineyard designated as well. The winery maintains 265 acres planted to the premier wine grapes: Chardonnay, Gewürztraminer, Sauvignon Blanc, Johannisberg Riesling, and Pinot Noir. This crop is supplemented with 50 acres at Firestone's Arroyo Perdido Vineyard planted to Merlot and Cabernet Sauvignon. (Both vineyards were planted under the direction of André Tchelistcheff, the great Californian oenologist, who is a consultant to Firestone.)

The modern winery is set against the Zaca mesa. From the air—a view handsomely depicted on some of its labels—it resembles a fan of cards, with the overlapping triangles of the roofline concealing dramatically different ceiling heights within, from the lofty fermenting room to the low-ceilinged (two barrels high at one end) aging cellars. All of these components are encased in a redwood and glass building (seen in a second series of labels)—all in all, a very dramatic structure.

The guided tour of the winery facilities is quite thorough. It includes a look at the Firestone vineyards, its crushing facility, barrel aging cellars, and other processes. (All winemaking operations are supervised by Alison Green, one of the few women winemakers in North America.) The tour ends with a stop in the wood-paneled tasting room whose walls are punctuated with windows revealing interior views of the dark aging cellars and exterior ones of the hills beyond the vineyards.

Firestone Vineyard makes only still table wine. It is particularly known for its Johannisberg Riesling and Gewürztraminer. Firestone labels are especially attractive and illustrate different aspects of vineyard and winery life or identify the different varietals. This artistic tradition was begun in 1976, and the line drawings on early bottles—created by the artist Sebastian Titus—have since become collector's items.

Open all year, except major holidays: Monday-Saturday, 10-4. Guided tours every 45 minutes (20 minutes). Tasting. Retail outlet: wine, gifts. Picnic facilities. Access for the handicapped. Directions: From Buellton, north on Highway 101. East on Zaca Station Road approximately 2½ miles to winery on left.

Zaca Mesa Winery
Foxen Canyon Road (P.O. Box 547)
Los Olivos, California 93441
Telephone: (805) 688-3310
Owners: Connie and Marshall Ream

Zaca Mesa Winery is located on a ranch in the Santa Ynez Valley of Santa Barbara County. Its annual output, 275,000 gallons, makes it second in the valley only to Firestone Vineyard, its neighbor on Zaca Station Road. The winery is in the middle of range land. Its vineyards, located on the Zaca mesa, consist of 250 acres planted to Sauvignon Blanc, Pinot Noir, Zinfandel, Chardonnay, Semillon, Merlot, Cabernet Sauvignon and Syrah.

At the bottom of the mesa lies the winery, housed in a rustic cedar-sided barn-like structure. Shaped like a horseshoe, the building contains the working winery, tasting room, and retail sales area. A particularly handsome courtyard for picnicking (the area enclosed by the horseshoe configuration) is shaded by massive oaks draped with Spanish moss.

Tours of the winery touch on its history and include a complete look at its operation, from crushing through fermentation, cooperage, and bottling. Afterwards, three to four wines are usually available for tasting. Sometimes a broader selection is available on weekends.

Zaca Mesa's product line features estate-bottled vintage-dated varietals. Since the winery's first vintage in 1975, its wines have been consistent award winners. Highly recommended are Reserve Pinot Noir, Reserve Sauvignon Blanc, and Zinfandel.

Open all year: daily 10-4. Guided tours. Tasting. Retail outlet: wine, gifts. Picnic facilities. Access for the handicapped. Directions: From Buellton, north on Highway 101 to Foxen Canyon Road. Follow road east for approximate-

*ly 9 miles, past Zaca Station Road and Firestone Vineyard, to winery which
will be on your left.*

Paso Robles

Estrella River Winery
Highway 46 East (P.O. Box 96)
Paso Robles, California 93446
Telephone: (805) 238-6300
Owners: The Cliff Giacobine Family

The relatively young family-owned-and-operated Estrella River Winery
is one of the pioneers in the up-and-coming Central Coast region. It is
the only winery in San Luis Obispo County that offers daily tours, and
it is also the largest in the Paso Robles appellation. Estrella was founded
in 1972 when Cliff Giacobine, a former aerospace engineer turned farmer,
planted 550 acres of a 1,000-acre farm he had purchased a year earlier to
premium wine grapes. In 1976, farm life apparently agreeing with him,
he planted an additional 160 acres, and in 1977 built his modern winery.
  Since then Cliff's wife, Sally, and their children have joined in the successful family enterprise. Estrella's vineyard has grown to 875 acres and
the winery's production to 70,000 cases. A pleasantly informal tour of
the Hacienda-style winery covers the vineyard, the winery itself (set on

a knoll amid the vines), the crush and fermentation facilities, barrel room, champagne production, and bottling line.

Estrella's wines are all estate bottled. Grapes grown in the surrounding vineyard include Chardonnay, Sauvignon Blanc, Chenin Blanc, Muscat Canelli, Johannisberg Riesling, Cabernet Sauvignon, Zinfandel, French Syrah, and Barbera. (The winery has the largest single planting of Syrah in the United States.) Of the varietal wines produced from these grapes, sixty-five percent are vintage dated. Ninety-eight percent is still table wine; the remaining two percent, a Blanc de Blanc *methode champenoise* sparkling wine.

A dozen picnic tables overlooking the vineyard, sheltered from the sun by a canopy, are available for lunch. More ambitious visitors can reserve in advance the use of a barbeque. Lunch meats, assorted cheeses, and fresh local bread can be purchased in the tasting room. An observation tower attached to the winery offers a spectacular view of the estate.

Open all year, daily except major holidays: 10-5. Guided tours: on the hour and half hour, 10-4:30. Free tasting. Retail outlet: wine, gifts, deli. Special events: Estrella Harvest Run, the county's most challenging race, first Sunday in November. Access for the handicapped. Directions: Highway 101 to Paso Robles. Winery 7 miles east of Paso Robles on Highway 46.

Temecula

Callaway Vineyard & Winery
32720 Rancho California Road
Temecula, California 92390
Telephone: (714) 676-4001
Owner: Hiram Walker & Sons, Inc.

A visit to Callaway, Southern California's largest premium winery, makes a good day's outing from either Los Angeles or San Diego, especially if you have reserved ahead for one of the firm's sumptuous barbeque lunches and special tours. The winery, founded in 1974 by Ely Callaway, is located in the rolling countryside four miles east of the Old West town of Temecula. Situated atop a hill surrounded by estate vineyards, Callaway's stucco and wood architecture is visible for some distance as you approach.

Regularly scheduled guided tours of this state-of-the-art facility last approximately twenty minutes. Visitors usually wait for their guide in the reception center, which contains a retail store selling both the Callaway product and a broad selection of wine-related gift items. From this vantage point you can also see the winery's huge tasting room, whose large windows overlook the Temecula Valley.

The standard tour begins outside with a view of Callaway's modern crush equipment and then proceeds into the winery for a look at the stainless-steel fermentation tanks, and the aging cellar, with its display of oak barrels and casks. Adjacent to the winery is the immense combined bottling and warehouse facility, whose glass-enclosed mezzanine allows you to watch the activity on the sterile floor below. The tour ends with a formal tasting and lecture, for which there is a charge of $1 per person (usually three wines are included).

While the regular tour is quite comprehensive, if you have the time, plan to make a day of it and sign up in advance for Callaway's in-depth tour, tasting, and barbeque luncheon. Held four or five times a week, this special tour begins at 11:00 with an hour-long visit to both the vineyards and the winery. At noon, an hour's tasting and lecture is scheduled, followed by a lavish buffet barbeque set out in a lovely grape arbor north of the main facility. The menu usually features chicken, beef sausage (both marinated in wine and barbequed over grape vines), Mexican specialties, corn on the cob, fresh fruit, salads, and more—all accompanied by a Callaway wine.

Callaway Vineyard makes only white wine. Its estate vineyards, planted to Chardonnay, Chenin Blanc, and Sauvignon Blanc, provide half of the grapes needed for its 110,000-case annual production; the remainder is purchased. The winery is known primarily for its premium white varietal dry table wines, but also produces a small quantity of Vin Blanc, a premium generic white.

Open all year, daily except major holidays: 10-5. Guided tours on the hour (last tour at 4:00). Tasting (with or without tour). Retail outlet: wine, gifts. Picnic facilities. Special events: In-depth tour, tasting, and barbeque lunch by reservation only, 4-5 times weekly, $20 per person; periodic seasonal "short courses" in winery procedure and viticultural technique, by reservation only. Access for the handicapped. Directions: From Interstate 15, take Rancho California Road east 4½ miles to winery entrance. Callaway is located approximately 60 miles north of San Diego and 90 miles southeast of Los Angeles.

# The Northwest

André Tchelistcheff, the world-famous California winemaker and consultant, has predicted that "the Northwest will eventually have a finer international reputation than California." Thus far this region, a relative viticultural youngster, is making great strides toward achieving such an exalted position in the world of wines.

Of the three Northwestern wine states of Washington, Oregon, and Idaho, Washington is currently the most important. Between 1968 and 1984 the number of wineries jumped from three to forty-seven. Grape-production in the early 1980s hovered around 159,000 tons, making Washington the second largest producer of grapes in the United States and its third largest producer of wine. With 9,000 acres planted to vinifera grapes since 1970, Washington is the second largest producer of quality wine grapes in America, and the state's potential for future growth is phenomenal. An estimated 150,000 acres of land are suitable for vineyards in the state, and when only one-third of this is planted, Washington's premium wine industry will equal that of the Napa/Sonoma region in size.

The center of Washington's wine industry is the metro-Seattle area. Close to the city center is Columbia Winery, one of the pioneers in the state's premium wine industry, and just around the corner is Washington's premier fruit winery, Paul Thomas. Twenty-five minutes north of the downtown area, in Woodinville, is the spectacular headquarters of Chateau Ste. Michelle, a U. S. Tobacco holding. The largest and most glamorous operation in Washington, Chateau Ste. Michelle has 3,000 acres planted to vineyards and two showcase wineries in addition to the Woodinville facility which offer some of the state's most agreeable wine-touring experiences and tastings. Each of the three wineries is richly furnished with European antiques and is set on lavishly landscaped grounds.

Washington's grapegrowing region also extends for 160 miles from the hills west of Yakima to the Columbia River, a semi-arid area that is being transformed to lush vineyards through modern scien-

224

tific irrigation. In addition to Chateau Ste. Michelle's two other wineries at Paterson and River Ridge, there are further wineries from which to choose, including elegant, German-owned F. W. Langguth Winery in Mattawa, and Preston Wine Cellars, the state's largest family-owned-and-operated winery, in Pasco. Further west in the Yakima Valley is very new, partially underground Quail Run, and tiny, choice, family-run Hinzerling Vineyards.

Oregon ranks second in the Northwest wine industry with thirty wineries, most of which are family-owned-and-operated. A majority of the wineries are found on the hillsides of the fertile Willamette Valley, but others are spread from Nehalem in the far north to Grant's Pass in the south, twenty miles from the California border. The greatest concentration of Oregon's wineries, including some of the state's best, is a short drive from Portland in Yamhill County. These include small Chateau Benoit Winery and Elk Cove Vineyards and two of the state's largest, Sokol Blosser Winery and Knudsen Erath Winery. The centrally-located town of McMinnville is a good base of operations if you want to make an overnight tour of these and the area's other interesting wineries.

Idaho is just emerging as a wine state. It has fewer than five wineries, all in the far western side of the state near the Oregon and Washington borders. Winemakers frequently buy grapes from nearby Washington's Columbia River Valley vineyards. The most important of Idaho's wineries, and one of the Northwest's largest, is Chateau Ste. Chapelle. A premium winery architecturally inspired by the Eglise de Ste. Chapelle in Paris, it is surrounded by estate vineyards near Sunny Slope and overlooks the rushing waters of the Snake River.

IDAHO

Caldwell

Ste. Chapelle
Highway 55 (Route 4, Box 775)
Caldwell, Idaho 83605
Telephone: (208) 888-9463
Owners: The Symms Family and Bill Broich

Although it takes fewer than the five digits on one hand to count Idaho's wineries, among its small number is Ste. Chapelle, one of the largest in the Northwest. Founded in 1971 by businessman, part-time apple grower, and amateur winemaker Bill Broich and his wife, Penny, the winery was inspired by the couple's tour of the wine country of France some years before. In addition to absorbing viticulture and vinification in the countryside, they traveled to Paris on weekends and were particularly captivated by the history and architecture of the Church of Ste. Chapelle. As a result of this vacation, Bill decided to become a full-time commercial winemaker and to name his winery, should it ever come to fruition, after the historic Paris church.

Five years later Broich established Ste. Chapelle, converting a farmhouse in Emmett for the initial crush. The grapes for his first wines were purchased from the neighboring Symms family, who had planted an experimental vinifera vineyard on their fruit ranch. Shortly thereafter Symms and Broich joined forces, expanded the vineyard acreage, and constructed Ste. Chapelle's present winery with the help of an architect armed with photographs of the French church. The winery is located on the Symms's ranch in Sunny Slope, about thirty-five miles from Boise and ten miles from Caldwell.

The winery and its vineyard have grown considerably since their inception. Ste. Chapelle now produces 150,000 cases of wine annually. Eighty-eight percent of its production is devoted to still table wines; the remainder, to *methode champenoise* and *charmat* (bulk) process sparkling wines. The vineyards at Sunny Slope now encompass 220 acres; 400 acres are also under contract with local growers (many just across the border in Washington State). Planted in the estate vineyard are Johannisberg Riesling, Chardonnay, Chenin Blanc, Gewürztraminer, Cabernet Sauvignon, and Pinot Noir. Ste. Chapelle's twelve wines are dominated by varietals, most of which are vintage-dated. (Some are vineyard-designated as well.) Ste. Chapelle is particularly known for its white wines, especially award-

226

winners Chardonnay and Johannisberg Riesling.

Short tours of the winery, which are given every day, include a look at its production facilities and aging cellar. Afterwards you can sample Ste. Chapelle's product in the tasting room or enjoy a picnic at one of the tables provided on the patio adjoining the winery. From both vantage points is a beautiful view of the extensive estate vineyard and orchards which slope down the hillside to the winding Snake River below; in the distance are Oregon's Owyhee Mountains.

Open all year: Monday-Saturday, 10-6; Sunday, 12-5. Guided tours. Free tasting. Retail outlet: wine, gifts. Jazz concerts (June-September: Sundays, 1-4:30). Picnicking on patio. Access for the handicapped. Directions: From Boise, Interstate 84 to Caldwell exit. South on Highway 55 to winery.

OREGON

Carlton

Chateau Benoit Winery
Mineral Springs Road (Route 1, Box 29B-1)
Carlton, Oregon 97111
Telephone: (503) 864-2991
Owners: Fred and Mary Benoit

One of Yamhill County's many wineries. Large concrete barn winery, on the crest of a hill, owned and operated by Fred and Mary Benoit. Adjacent young twenty-two acre vineyard is planted to Riesling and Müller-Thurgau. From these grapes and the harvest of the Benoits' older vineyard in Lane County, planted in 1972 to Chardonnay, Pinot Noir, and Riesling, UC-Davis-trained winemaker Rich Cushman produces 26,000 gallons annually. The wines, consistent award-winners, are Swiss-German in style and slightly sweet. Labels include varietal wines plus Brut and Blanc de Blancs methode champenoise sparkling wines, Pinot Noir Nouveau, Sauvignon Blanc, and Merlot. Tasting area in a corner of the working winery. Open all year: Monday-Friday, 11-5; Saturday and Sunday, 12-5. Guided tours by appointment. Free tasting (usually 6 or 7 wines available). Retail outlet: wine, gifts. Bastille Day Festival (July). Picnic facilities. Access for the handicapped. Directions: From McMinnville, 99W northeast towards Portland. Turn left on Mineral Springs Road. Winery 1½ miles on your right.

Cave Junction

Siskiyou Vineyards
6220 Oregon Caves Highway (Highway 46)
Cave Junction, Oregon 97523
Telephone: (503) 592-3727
Owner: C. J. David

Siskiyou, Oregon's southernmost winery, is located at the edge of the Siskiyou National Forest in the state's scenic Illinois Valley. Less than twenty miles from the California border and only six from the gateway to the Oregon Caves National Monument, the winery is a convenient stop for the estimated 120,000 tourists who visit the area annually.

Siskiyou Vineyards was bonded in 1978 by Charles and Suzi David on their 100-acre ranch along Bear Creek. The Davids had not planned to establish a vineyard or start a winery—Charles is an electronics engineer by training—but in 1974, when they purchased the property, they discovered a stack of grape vines in one of the outbuildings and decided to plant them.

As a result, the Davids now have a thriving twelve-acre vineyard planted to Merlot, Cabernet Sauvignon, Gamay Beaujolais, Pinot Noir, Gewürztraminer, and Chenin Blanc; a new chalet-style winery and tasting room; and a new venture—Bear Creek Supply Company, which offers advice to others interested in viticulture.

Tours of the small, attractive winery's production facilities are by appointment only. But the tasting/retail area is open daily for more impromptu visits. Siskiyou produces 6,000 cases of still table wines annually, both varietals and generics. All wines are made from southern Oregon fruit (about two-thirds of the grapes needed for production are purchased). Many of the varietals, including the '82 Chardonnay and Johannisberg Riesling, have won awards.

Open all year: daily, 11-5. Guided tours by appointment only. Free tasting. Retail outlet: wine, gifts. Festivals: Southern Oregon Wine Festival (June); Ladies Softball Tournament (Summer); Bike Run (Fall). Picnic facilities. Access for the handicapped limited to ground floor. Directions: Redwood Highway 199 north to Cave Junction. At Cave Junction take Highway 46 east, 6.2 miles to winery.

Dundee

Knudsen Erath Winery
Worden Hill Road (Route 1, Box 368)
Dundee, Oregon 97115
Telephone: (503) 538-3318
Owners: Richard Erath and Cal Knudsen

Knudsen Erath, Oregon's largest winery, is one of many located in Yamhill County, where varietal grapes thrive in the reddish-orange clay soil of the area's famous Red Hills of Dundee.

Established in 1969, Knudsen Erath is the joint effort of Cal Knudsen and Dick Erath. Born and raised in California, Erath comes from a line of German winemakers (including both his father and grandfather). He bought his original vineyard acreage from Knudsen; the two later joined forces to establish the winery.

As you approach Knudsen Erath, you'll first see its vineyard, since the vines are planted right down to the edge of the entrance road. Totaling 100 acres, the vineyard is planted to equal numbers of Pinot Noir, Riesling, and Chardonnay. (Erath's original vineyard also contains small plantings of Merlot, Gewürztraminer, and Gamay Beaujolais.) From these estate-grown grapes and small amounts provided by local growers, the winery currently produces 51,000 gallons of wine annually. Ninety-five percent of its product line is devoted to table wines, both varietals and proprietary blends (the latter bottled under the "Dundee Villages" label). The balance is given to *methode champenoise* sparkling wine. The varietal Pinot Noir has won numerous awards.

The handsome cedar-sided, shake-roofed hospitality center is situated on a high promontory, surrounded by venerable oak trees shrouded with mistletoe. To the east is the Cascade Mountain range. On clear days Mt. Hood and Mt. Jefferson are visible in the distance. Down the hill from the visitor center is the similarly designed production facility. Tours are given by appointment only and take approximately an hour. They begin in the vineyard with an explanation of pruning and cultivation, and cover the production equipment and aging room as well. Hospitable tastings include bread and cheese. While no picnicking is allowed on the property, you might like to take a hamper to nearby Crabtree Park.

Open all year: Monday-Friday, 10-3; Saturday and Sunday, 11-5. Guided tours by appointment only, 12-4. Free tasting. Retail outlet: wine, gifts. Annual Harvest Festival. Access for the handicapped. Directions: From Highway

99 West in downtown Dundee, go west on 9th Street (which becomes Worden Hill Road) 2½ miles to winery. Winery road is first left after Crabtree Park turnoff.

Sokol Blosser Winery
Blanchard Lane (P.O. Box 199)
Dundee, Oregon 97115
Telephone: (503) 864-3342
Owners: The Sokol and Blosser Families

Sokol Blosser, another of Yamhill County's many wineries, is not only one of Oregon's largest, but one of its most handsome. Its combined tasting room and underground aging cellar was designed by one of the state's leading architects, John Storrs; it is surrounded by orchards and vineyards, and the hilly setting offers sweeping views of the Willamette Valley.

The Sokol and Blosser families own and operate the winery. Susan Sokol Blosser, a history professor, manages the vineyard, while her husband, Bill, a Portland land-use planner, tends the winery. Their winemaker is Dr. Robert McRitchie, a biochemist and microbiologist, who gained his winemaking experience in the Napa Valley.

The winery opened in 1977, six years after the initial forty-acre vineyard was planted. It now produces 22,000 cases of still table wine annually. Currently forty-five acres of vineyards are under cultivation, planted to Pinot Noir, Chardonnay, Johannisberg Riesling, and Müller-Thurgau. These producing vineyards provide half of the winery's grapes; the rest are purchased from local growers. Sokol Blosser's product list includes both varietal and generic wines. Its best wines carry the appellation Yamhill County; exceptional ones carry vineyard designations as well, such as Red Hill and Hyland Vineyards. All wines are vintage-dated. Sokol Blosser's Pinot Noir, Chardonnay, and Johannisberg Riesling are especially noteworthy.

Guided tours of the winery (located in a separate building from the aging cellar and tasting room) are offered by appointment only. But if you happen to stop by unexpectedly for a complimentary tasting, you can look into the winery from a special viewing platform.

Open all year: daily, 11-5. Guided tours, $1, by appointment (one day's notice required). Free tasting. Retail outlet: wine, gifts. Festival: Wine Country Thanksgiving (three days after Thanksgiving). Picnic facilities. Access for the handicapped. Directions: From Dundee, 2½ miles southwest on Highway 99W to Blanchard Lane. Right on Blanchard Lane to winery.

Forest Grove

Tualatin Vineyards
Seavy Road (Route 1, Box 339)
Forest Grove, Oregon 97116
Telephone: (503) 357-5005
Owners: William H. Malkmus and William Fuller

Large modern winery, founded in 1973, is one of the three biggest in the state. Located in the foothills of Oregon's Coast Range. Eighty-five acres planted to Chardonnay, Johannisberg Riesling, Gewürztraminer, Muscat, and Pinot Noir. Winemaker and part-owner William Fuller educated at UC-Davis and instructed by Louis M. Martini. Production: 18,000 cases of still table wine annually, 100 percent varietals, vintage-dated and estate-bottled. Best known for Pinot Noir and Chardonnay, the latter voted "best in the world" in recent European wine competition. Scenic picnic facilities in a park-like setting overlooking the Willamette Valley. Open all year, except January: Saturday and Sunday, 1-5. Guided tours. Fourth of July Barrel Tasting, Thanksgiving Art Show. Picnic facilities. Access for the handicapped. Directions: From Portland, Sunset Highway (US 26) west, 30 miles to State Road 6. State Road 6 to 47. 47 south to Clapshaw Hill Road; west on Clapshaw Hill Road to Seavy Road and winery. (There are signs to the winery from State Road 6 turn-off.)

Gaston

Elk Cove Vineyards
Olson Road (Route 3, Box 23)
Gaston, Oregon 97119
Telephone: (503) 985-7760
Owners: Patricia A. and Joe H. Campbell

Down the road from Tualatin Vineyards is Elk Cove, a small premium winery owned and operated by physician Joe Campbell, his wife, Pat, and their five children. The couple, largely self-taught winemakers and viticulturists, established their vineyard and winery in the foothills of the Oregon Coast Range west of Portland, naming it for the Roosevelt elk which migrate to the Willamette Valley each spring.

Since their first plantings in 1974, the Campbells' vineyard has grown to twenty-four acres of Pinot Noir, Chardonnay, Gewürztraminer, and Johannisberg Riesling. From these grapes and additional purchased fruit, notably Cabernet Sauvignon, the Campbells produce 30,000 gallons of wine annually, ninety-five percent of which is vintage-dated and vineyard-designated. Elk Cove is particularly known for its Chardonnay, Johannisberg Riesling, and Pinot Noir.

Short guided tours of Elk Cove's new winery building, situated amid a sheltering grove of pines, are given year round by appointment. You'll appreciate the breathtaking view from the tasting room's windows as you sip the wine. To the east you can see the entire vineyard; to the west are the mountains of the Coast Range.

Open all year, except Christmas and Thanksgiving: daily, 12-5. Guided tours by appointment (one week's notice suggested). Free tasting. Retail outlet: wine, gifts. Riesling Festival (Memorial Day weekend) and annual Pinot Noir Picnic. Picnic facilities. Access for the handicapped. Directions: From McMinnville, Highway 47 north 20 miles to Olson Road (just before town of Gaston). West on Olson Road, 2.8 miles to winery.)

Hillsboro

Oak Knoll Winery
Burkhalter Road (Route 6, Box 184)
Hillsboro, Oregon 97123
Telephone: (503) 648-8198
Owners: Ron and Marjorie Vuylsteke

Founded in 1970, winery is located just outside Hillsboro village center in the Willamette Valley. Produces award-winning fruit and berry wines from native crops as well as premium varietal wines from viniferas. Especially noted for Raspberry and Pinot Noir. Family owned and operated. Most grapes purchased. Open all year: Wednesday, Thursday, Friday, and Sunday, 12-5; Saturday, 11-5. (Open Monday and Tuesday by appointment.) Self-guided tours. Guided tours by appointment. Free tasting. Retail outlet: wine, gifts. Bacchus Goes Bluegrass Festival (mid-May). Case Sale (weekends before and after Thanksgiving). Picnic facilities. Limited access for the handicapped. Directions: From Portland, Tualatin Valley Highway (8) west to Hillsboro. South from town center on Highway 219 4 miles to Burkhalter Road. Burkhalter Road east to winery.

Roseburg

Hillcrest Vineyard
240 Vineyard Lane
Roseburg, Oregon 97470
Telephone: (503) 673-3709
Owner: Richard H. Sommer

Richard Sommer, owner of Hillcrest Vineyard, is considered a pioneering member of Oregon's modern winegrowing industry. An agronomy graduate of UC-Davis, Sommer came to the state in the late 1950s in search of a cool region in which to grow German-style wine grapes. Choosing a gentle knoll in the Umpqua River Valley of southern Oregon, he planted the state's first vinifera vineyard (five acres) with cuttings from the Napa Valley.

Today, Sommer's vineyard encompasses thirty-five acres, predominately planted to Johannisberg Riesling, Gewürztraminer, Pinot Noir, and Cabernet Sauvignon, the last of which is rarely grown in Oregon. His vineyard produces 23,000 gallons of wine annually and is particularly noted for estate-bottled vintage-dated varietal wines, especially its award-winning Johannisberg Riesling. Hillcrest's Pinot Noir and Cabernet Sauvignon have won a number of awards, as well.

Guided tours of the rustic winery, which is set in the middle of the vineyard, are given by knowledgeable staff members who tailor their remarks to the interest of the visitors. In the pine-paneled tasting room, seven or eight Hillcrest wines are usually available for sampling. There are tables for picnicking indoors, or you may choose to sit outside on a deck overlooking the vineyard.

Open all year, except major holidays: daily, 10-5. Guided tours. Free tasting. Retail outlet: wine. Picnic facilities. Access for the handicapped. Directions: Interstate 5 to Garden Valley Road. Northwest on Garden Valley Road to Melrose Road. West on Melrose for 3 miles to Cleveland Hill Road. North on Cleveland Hill to Orchard Lane, west on Orchard Lane to Elgarose Road. Follow Elgarose to Vineyard Lane and winery. (There are signs marking the route from Melrose Road.)

WASHINGTON

Bellevue

Columbia Winery
1445 120th Avenue N.E.
Bellevue, Washington 98005
Telephone: (206) 453-1977
President: Daniel R. Baty

Columbia Winery, formerly called Associated Vintners, occupies a functional warehouse north of Lake Bellevue, just a ten-minute drive from downtown Seattle. Its plain exterior belies its special attributes, however. It is the oldest premium winery in the state of Washington and is considered a pioneer in the production of great European varietals. Columbia's winemaker, David Lake, is the only active Master of Wine (a prestigious British accreditation) in this country. Lake trains all of the tour guides, who must learn the workings of both vineyard and production facility.

Columbia Winery was founded in 1962 by ten wine enthusiasts from the University of Washington. Since its first crush in 1967, completed in a Seattle garage, the winery has moved three times while growing to its current output of 100,000 gallons annually. Muscat, a stray feline who adopted the vintners and has since become their mascot, oversees production. The winery's product list includes Semillon, a long-standing specialty; Cabernet Sauvignon; Gewürztraminer; Chardonnay; Johannisberg Riesling, both the traditional dry and a multi-award-winning semi-dry labeled "Cellarmaster Reserve"; Merlot; and Pinot Noir. In 1984 alone, Columbia's wines were awarded no fewer than twenty-four honors. All grapes are purchased from Washington growers (under the current stewardship of Dan Baty, the winery divested itself of its vineyard holdings to focus on winemaking). All wines produced are vintage-dated; many are vineyard-designated as well.

Guided tours of the winery take anywhere from twenty minutes to an hour depending on the questions and interests of the group. You then can sample Columbia's wines in its newly remodeled tasting room, which is decorated with ceramic tiles and photographs of the Yakima Valley and Columbia Valley vineyards which supply its fruit. Five wines can be tasted daily; the selection changes every week.

Columbia Winery sponsors numerous festive events in connection with its seasonal activities; check ahead for details.

*Open all year: Monday-Saturday, 8:30-4:30; Sundays (May-October only),
11-4. Free tasting. Retail outlet: wine, wine accessories, books. Picnic facilities
planned. Many special events. Call or write winery for current schedule. Access for the handicapped. Directions: From Seattle, take the Evergreen Bridge
(520) eastbound past the 405 Interchange. Exit at 124th Avenue N.E., then
west on Northrup Way to 120th Avenue N.E. Turn south on 120th to winery
(on your right before Lake Bellevue).*

Paul Thomas Wines
1717 136th Place, N.E.
Bellevue, Washington 98005
Telephone: (206) 747-1008
Owners: Paul and Judy Thomas

*Most sophisticated fruit and berry winery in the state (on a par with Oak
Knoll in Oregon); only a few blocks east of Columbia Winery. Also one of
the larger producers of wine in Washington (33,000 gallons annually). Family
owned and operated. Established its reputation with Crimson Rhubarb, a
best seller. Also produces Dry Bartlett Pear and Washington Raspberry.
Recently, product line has been expanded to include vinifera wines, including
Johannisberg Riesling, Sauvignon Blanc, Cabernet Sauvignon, and Muscat
Canelli. Dignified, elegantly furnished tasting room reflects the high quality of the wines produced. Open for tours and tastings by appointment only.
Retail sales: Monday-Friday, 9-5. Access for the handicapped. Directions: From
Seattle take Evergreen Bridge (520) eastbound past the 405 Interchange. Ex-
it at 124th Avenue N.E. East on Northup Way to 136th Place, N.E. South
on 136th to winery (on your right).*

Mattawa

F. W. Langguth Winery
2340 South West Road F-5
Mattawa, Washington 99344
Telephone: (509) 932-4943
General Manager: Max Zellweger

Bonded in 1982, F. W. Langguth is one of Washington's youngest wineries,
but it is already one of the largest. The winery was founded by German
wine magnate Johann Wolfgang Langguth, owner of one of the Moselle

Valley's largest wine holdings. Located in the Columbia River Valley in southeastern Washington, the 35,000-square-foot winery is a modern, state-of-the-art facility that fully implements the latest northern European technology. It has a tank capacity of 500,000 gallons, 265 acres of estate vineyards, and a Swiss- and German-trained winemaker, Max Zellweger.

From the surrounding vineyards planted to Johannisberg Riesling, Chardonnay, and Gewürztraminer (and from additional purchased grapes) winemaker Zellweger produces primarily varietal table wines, with ten percent of the production given to proprietary reds and whites. All wines are vintage-dated; the winery is particularly known for its Rieslings, especially the Late Harvest.

Short guided tours include a look at the crushing area, fermentation tanks, bottling line, and warehouse. Tours conclude in the winery's elegant tasting room. Reflecting the firm's old world roots, it is finished with stained-glass windows and crystal chandeliers. White linen cloths cover the indoor tables where you can sample the wine and enjoy a picnic lunch. There are additional picnic facilities outdoors, where panoramic views of the distant mountains, including snow-capped Mount Rainier to the west, can be enjoyed.

Open all year: daily, 10-5. Guided tours. Free tasting. Retail outlet: wine, gifts. Annual Weinfest (July). Picnic facilities. Access for the handicapped. Directions: From Seattle, Interstate 90 to Highway 243 (Vantage). South on 243 to Mattawa. At Mattawa follow South West Road 24 (for 13 miles) to Road F-5 and winery. There are signs directing you from Mattawa.

Pasco

Preston Wine Cellars
502 East Vineyard Drive (Star Route, Box 1234)
Pasco, Washington 99301
Telephone: (509) 545-1901
Owners: Bill and Joann Preston

If you travel northeast from F. W. Langguth Winery you will come to Preston Wine Cellars, the largest of Washington State's family-owned wineries. Bill Preston, once a salesman of farm irrigation equipment, was well-versed in the special watering techniques needed to nurture a vineyard in the fertile, yet arid, farmlands of Washington's Southern Columbia Basin. Armed with this knowledge, he bought the 181 acres where his winery and vineyard now stand in the early 1970s and began to plant an

all-vinifera vineyard on 50 of them, installing a computer-controlled overhead sprinkling system to sustain his vines. His vineyard has since grown to 130 acres and is planted to eleven vinifera varieties, Chardonnay, Sauvignon Blanc, Gewürztraminer, Johannisberg Riesling, Chenin Blanc, Pinot Noir, Merlot, Cabernet Sauvignon, and Gamay Beaujolais among them.

In 1976, the Prestons broke ground for their winery. It currently produces 130,000 gallons annually, devoted almost exclusively to estate-bottled varietal wines and to two generics, Desert Gold and Desert Rose. Preston's wines have consistently won awards since the first crush. Particularly noteworthy are the Fumé Blanc and Chardonnay. Bill Preston gives much of the credit for his well-received wines to the land and the high-quality grapes it produces. He believes that this part of the state is destined to become the white wine capital of the world.

The winery, featuring an attractive chalet-style tasting room on its roof, has become a popular tourist attraction, with as many as 1,000 visitors on a given weekend. Tours are self-guided; explanatory signs are posted along catwalks above the working winery. A ramp leads to the upstairs tasting room, where you can sample the Preston product and browse through the wide selection of wine-related gift items on display. The two-story tasting area, finished with natural woods, incorporates seating for sixty on hand-crafted furniture carved with grape motifs. Permanent exhibits of corkscrews, wine bottles, and corks decorate the walls. A large exterior deck affords views of the Preston vineyard. A picnic area is located in what the Prestons call the "conversation-piece park," a lovely spread of verdant lawn where an eclectic array of local artifacts, including an antique steam engine, is displayed.

Open all year, except major holidays: daily, 10-5:30. Self-guided tours. Free tasting. Retail outlet: wine and gifts. Open House (July). Picnic facilities. Access for the handicapped. Directions: 5 miles north of Pasco on 395 (Spokane Highway). (Watch for sign on east side of highway.)

Prosser

Hinzerling Vineyards
1520 Sheridan Avenue
Prosser, Washington 99350
Telephone: (509) 786-2163
Owners: The Wallace Family

Small premium estate winery in southeastern Washington's Yakima Valley, owned and operated by Wallace family (Mike Wallace is one of Washington's winegrowing pioneers). Twenty-five acre vineyard, established in 1972, planted to Cabernet Sauvignon, Merlot, Malbec, Cabernet Franc, Johannisberg Riesling, and Gewürztraminer. Winery, in unimposing cinderblock structure, bonded in 1976. Product line primarily estate-grown vintage-dated varietals. One generic blend: Ashfall White. Best regarded for Late Harvest and Botrytis wines from Riesling and Gewürztraminer, and Ashfall White. (Hinzerling is the state's foremost producer of botrytised wine and one of the few that makes sparkling wine.) Winery named for German family that settled the area; Ashfall White named for the ashes Mt. Saint Helens recently spewed across the state. Open all year. Monday-Saturday: 10-12, 1-5. Sundays: April-November 12-4. Guided tours by appointment only (24 hours notice suggested). Free tasting. Retail outlet: wine, gifts. Barrel Tasting (spring); Open House (Fourth of July); Tasting the New Vintage (fall). Access for the handicapped. Directions: Interstate 82 to Prosser, exit 80. Proceed into city on Highway 12, turning left on Sheridan Avenue. Winery is at end of Sheridan.

Woodinville

Chateau Ste. Michelle
One Stimson Lane (P.O. Box 1976)
Woodinville, Washington 98072
Telephone: (206) 488-1133
Owner: United States Tobacco Company

Chateau Ste. Michelle is certainly Washington State's preeminent winery and is an industry leader in the Northwest as a whole. Its statewide holdings include more than 3,000 acres of vineyards and three showplace wineries at Woodinville, Paterson, and Grandview. (The Woodinville Chateau, twenty-five minutes north of Seattle, is the headquarters for the company.)

From these impressive facilities and extensive vineyards planted to Johannisberg Riesling, Chardonnay, Sauvignon Blanc, Chenin Blanc, Cabernet Sauvignon, Merlot, Pinot Noir, Grenaché, and Semillon, Chateau Ste. Michelle produces over half a million gallons of 100 percent varietal wine annually. The product list, with more than a dozen varieties, includes both *methode champenoise* sparkling wines and still table wines. Ste. Michelle's team of winemakers, advised by noted authority André Tchelistcheff, produces some of the best wines the Northwest region has to offer. Although the list of awards is a long one and almost every varietal has been honored, Ste. Michelle is particularly noted for Johannisberg Riesling, Chardonnay, Cabernet Sauvignon, and Merlot.

Each of Chateau Ste. Michelle's three wineries has extensive visitor facilities; all are counted among the state's biggest tourist attractions. Chateau Ste. Michelle in Woodinville is the most elaborate of the three and the best known, since it was the first opened to the public. Its winery and visitor reception area are located in a formal French chateau on the eighty-seven-acre Stimson estate amid park-like grounds with manicured lawns, elaborate gardens and ponds, experimental vineyards, and even a trout-stocked lake. Comprehensive cellar tours are given daily, after which you can sample Chateau Ste. Michelle's wines in a handsome, antique-furnished tasting room. The adjoining wine shop not only carries Ste. Michelle wines, but picnic supplies to enjoy at one of the numerous picnic sites on the beautiful grounds.

In eastern Washington, at the heart of Ste. Michelle's primary vineyards, is its recently completed River Ridge Winery in Paterson. Designed in the style of a French manor house, its massive redwood and stone visitor center is richly furnished with European antiques, a stained-glass mural, and replicas of ancient French tapestries. Tastefully integrated into the decor of the reception area are various educational exhibits displayed in restored antique cabinets. These describe the history and making of Chateau Ste. Michelle wines and explain their relation to the classic wines of Europe. On the wall a three-dimensional relief map illustrates the unique viticultural conditions of the Columbia Valley. Like the Woodinville facility, the River Ridge Winery offers guided tours of its cellars followed by a free tasting. A well-stocked retail shop carries both picnic fare and wine. The grounds offer splendid views of the Columbia Valley and the estate vineyards.

Ste. Michelle's third winery is located outside the rural town of Grandview amid its Yakima Valley vineyards. Grandview, recently opened to visitors, is considerably smaller and less elaborate than its Woodinville and River Ridge counterparts, but it, too, offers a tour, a free tasting, and a retail outlet with wine and picnic supplies. The Grandview winery, producing wine since the 1930s, is the state's oldest winemaking facility. Since 1978, Chateau Ste. Michelle has fermented and blended all of its red wines here.

Any of Chateau Ste. Michelle's three holdings is certainly worth including on any wine lover's itinerary. The Woodinville and River Ridge properties host numerous civic and non-profit events each year that make visits to these locations even more attractive. Such events include concerts, art exhibits, and flower shows.

Woodinville: *Open all year: daily, 10-4:30. Guided tours. Free tasting. Retail outlet: wine, gifts, picnic supplies. Special events. Picnic facilities. Access for the handicapped. Directions: From Seattle, take Interstate 405 north to exit 23B. Exit onto 124th Avenue heading east. At Old Woodinville Highway, turn left and proceed to 145th Avenue. Left on 145th to winery. (Winery is 15 miles northeast of Seattle.)*

Paterson: *Open all year: daily, 10-4:30. Guided tours. Free tasting. Retail outlet: wine, gifts, picnic supplies. Special events. Picnic facilities. Access for the handicapped. Directions: 35 miles southwest of Tri-Cities, on Highway 221. Winery is 1 mile north of intersections of Highways 221 and 14.*

Grandview: *Open all year: Tuesday-Sunday, 10-5. Guided tours. Free tasting. Retail outlet: wine, gifts, picnic supplies. Limited access for the handicapped. Directions: Interstate 82 to Grandview exit. Winery is in center of town at corner of West Fifth and Avenue B.*

Zillah

Quail Run Vintners
Morris Road (Route 2, Box 2287)
Zillah, Washington 98953
Telephone: (509) 829-6235
General Manager: Stan Clarke

Quail Run, one of the Yakima Valley's newest wineries, is also one of its most attractive. Its thoroughly modern winery, bonded in 1982, is located above the town of Zillah in the foothills of the Rattlesnake Mountains. The 6,000-square-foot structure, constructed of concrete and cedar, utilizes a true cellar concept in its design. Extensive earth berming around the production area and a clerestory above provide cooling by both day and night and keep the interior temperature at an optimum level for the maturing wine. The tasting room is located on the mezzanine level and overlooks the vineyards and orchards of the Yakima Valley. On another wall, interior insulated glass provides a view into the cellar. Doors from the tasting room lead to a spacious patio where you can enjoy the wine, have lunch, or just admire the wonderful scenery.

Quail Run was founded in 1981 by two independent groups of apple growers and wine enthusiasts. Spearheaded by UC-Davis-trained manager Stan Clarke and winemaker Wayne Marcil, its first vintage was in 1982. Just over 30,000 gallons were produced, but the wines in this first lot received sixteen medals in competitions throughout the United States, including a Best of Show for the Johannisberg Riesling. Currently, Quail Run has 100 acres under cultivation planted to Johannisberg Riesling, Chardonnay, Gewürztraminer, Merlot, Cabernet Sauvignon, Semillon, and Muscat Blanc. From this harvest and additional purchased grapes, its output of table wine has risen to 80,000 gallons, with expectations that it will reach 100,000 in the near future. The Rieslings, which are vintage-dated, are currently Quail Run's best wines; the Cabernet and Merlot have yet to be released.

Open all year: Monday-Saturday, 10-5; Sunday, 12-5. Guided tours by appointment only (one day's notice suggested). Free tasting. Retail outlet: wine, gifts. Picnic facilities. Access for the handicapped. Directions: From Yakima, take Interstate 82 southeast (approximately 20 miles) to exit 52. Proceed towards Zillah; turn left onto Fifth Street. Follow Fifth for 2 miles (it becomes Rosa Drive). Turn right at stop sign on Highland Drive. Follow Highland for 1 mile to Morris Road. Turn left on Morris and follow it 1½ miles to winery, which is on your left.

CANADIAN WINERIES
AND VINEYARDS

 Canada

Canada's wine history predates that of most of the United States, yet its post-Prohibition recovery has been slower, hampered in large part by the country's liquor store monopoly system imposed at Repeal. Grapes were first fermented on Canadian shores by Jesuit missionaries using the wild stock that grew prolifically along the banks of the St. Lawrence River in the 1600s. Two-hundred years later the Dominion's first commercial winery was established near Lake Ontario. And by the late 1800s, 5,000 acres of land were planted to wine grapes in the province of Ontario and thirteen wineries had been founded.

But in the early 1900s Canada, too, was swept along by the Prohibitionist spirit which spread north from the United States, and all of the provinces voted themselves dry, with the exception of Quebec, inhabited by the wine-loving French. In 1927 the principal grapegrowing province of Ontario voted for repeal (six years earlier than the United States), and for a short time the wineries enjoyed a heyday exporting wine across the border to American bootleggers.

Until the California wine revolution of the 1960s caught up with Canada in the early '70s, there was little softening of the laws restricting distribution and the corresponding growth of the wine industry. In 1975 the first new Ontario winery license since 1929 was issued to Inniskillin Wines, and, two years later, Canada's wineries were permitted to open their doors to the public for tastings and tours.

Although a relative latecomer to the delightful world of touring, Canada offers some attractive possibilities for the traveling wine lover, especially in the provinces of Ontario and British Columbia, which encompass the country's main grapegrowing regions. Many of the wineries of Quebec are closed to the public, and most depend primarily on imported processed juice for their operations.

Ontario, Canada's oldest and largest grapegrowing region and its most fertile, has the greatest number of wineries and, consequently, is more prepared for visitors than other provinces. Here on the flat farmlands of the Niagara peninsula, with a climate moderated by the waters of Lakes Ontario and Erie, are many of the nation's largest and most historic wineries, including Brights and Barnes near Niagara Falls and small cottage (farm) wineries such as Inniskillin and Reis. On the southern reaches of the peninsula is the new Pelee Island winery with its offshore vinifera vineyard—the largest premium vineyard in Canada—and in nearby Harrow is found the first European-financed winery on Canadian soil, the Italian-owned Colio.

Second to Ontario in vinous importance is the new viticultural region of the Okanagan Valley of British Columbia, where the local government began licensing farm wineries in the early 1970s. Although a dry, arid, desert-like region, the area's climate is moderated by the deep waters of Lakes Okanagan and Skaha. These lakes also provide water to irrigate the vineyards planted on surrounding hillsides. Among the most interesting operations to be visited is Brights House of Wines located on the Osoyoo Indian Reservation. Here much experimentation with unusual grape varieties is being conducted. Also not to be missed is Mission Hill's beautiful California-style winery at Westbank with its large picnic lawn, or popular Casabello Wines in Penticton which incorporates a film on winemaking and viticulture in its guided tour. In addition, there is also a wealth of new, small premium estate wineries to round out your visit to the area, including Claremont Wines and Gray Monk Cellars.

No discussion of Canada's wineries would be complete without a mention of its newest and smallest viticultural region, the maritime province of Nova Scotia. Although winegrowing is just beginning here, progress has been rapid. The southwestern side of the island has proved to be especially hospitable to vineyards as the climate here is moderated by the Bay of Fundy. The province now has two cottage wineries. Provincial laws regarding the sale of liquor, however, have yet to catch up with modern times. The wineries are not permitted to sell or offer tastings of their vintages. Tours, nevertheless, offer the visitor a rewarding experience.

BRITISH COLUMBIA

Okanagan Centre

Gray Monk Cellars
Camp Road (P.O. Box 63)
Okanagan Centre, British Columbia V0H 1P0
Telephone: (604) 766-3168
Owner: George Heiss

Gray Monk Cellars, located in British Columbia's northern Okanagan Valley, is one of the attractive small wineries begun in the region when the provincial government began licensing estate wineries in the early 1970s. Sometimes called "cottage wineries," these enterprises are permitted to sell up to 30,000 gallons of wine annually, all of which must be made from British Columbia grapes (fifty percent of it estate-grown).

Apple growers George and Trudy Heiss established Gray Monk's vineyards in 1972, replanting their orchards to the European grape varieties Johannisberg Riesling, Pinot Gris, Bacchus, Keiner, Maréchal Foch, Gewürztraminer, and Ehrenfilfer. From the harvest of their thirty-three-acre vineyard and additional purchased fruit, the Heisses produce 100-percent varietal wine; they are particularly known for their prize-winning Pinot Gris and Bacchus.

Tours of Gray Monk Cellars begin on the deck which wraps around three sides of the Spanish-modern winery building. Your visit will include the crushing area, fermentation tanks, and bottling line; usually all eight Gray Monk varietals may be sampled in the tasting room. You'll probably want to spend some time on the winery deck where picnic tables are available. Gray Monk's hilly location affords lovely views of the vineyards, which cover the slope down to Okanagan Lake; mountains loom in the distance.

Open all year: Tuesday-Saturday. Summer, 10-5; Winter, 1-5. Guided tours: 11-4. Free tasting (with or without tour). Retail outlet: wine. Picnic facilities (tables on deck and in courtyard). Access for the handicapped. Directions: North on Highway 97 to Winfield. (Winery six kilometers west of Winfield; signs mark the way from Winfield.)

Oliver

Brights House of Wine
Highway 97 (P.O. Box 1650)
Oliver, British Columbia V0H 1T0
Telephone: (604) 498-4981
Owners: T. G. Bright and Co., Ltd.

Brights House of Wine is the new British Columbian branch of Brights Wines, one of Canada's leading winemakers headquartered in Ontario; it is also one of the newest commercial wineries to open in this province. Its spectacular facility, completed in 1982, is situated in the semi-arid area of the Monashee Mountain range on land leased from the local Osoyoos Indian tribe. The winery building is owned by the Osoyoos, who built it with funds provided by the federal and provincial governments. In a joint venture with the tribe, Brights has leased the building for five years, equipping it with the expensive machinery needed for wine production. The Osoyoos tend the vineyards, which extend beyond the winery and now comprise 250 of their 6,000 tribal acres.

   Brights has implemented an extensive research program to evaluate new grape varieties; thirty-five different ones from Russia, Hungary, Spain, Germany, and France are planted here, many of which are nowhere else in North America. The more promising of the wines produced are made available in the winery's retail store, but quantities are limited and sell out quickly (you'll find the best selection in July and August).

   Tours of Brights House of Wine take about an hour, including the tasting. Guides will show you the crush area, cellar, bottling line, and other production facilities. Four wines can usually be sampled in the cedar-paneled tasting room.

   Brights' British Columbia facility, with a 750,000-gallon storage capacity, makes table, sparkling, and fortified wines. Experimental vintages are sold under the Vaseaux Cellars label; generics are sold under the name Brights Wines; and popular "bag-in-box" wines are available in several sizes. The company is noted for its Baco Noir and limited-edition Matsvani, which is produced from a Russian grape variety.

Open May-September: Monday-Sunday, 10-6 (rest of year by appointment only; one day's notice required). Guided tours. Free tasting (with or without tour). Retail outlet: wine, gifts. Wine Festival (first week in October). Picnic tables on spacious grounds. Access for the handicapped limited to tasting room and retail outlet. Directions: Take Highway 97 from Oliver; winery is 5 miles north.

Peachland

Claremont Estate Winery
Trepanier Bench Road (S26 C4 R.R. 2)
Peachland, British Columbia V0H 1X0
Telephone: (604) 767-2992
Owners: Bob and Lee Claremont

Claremont Estate Winery, opened in 1979, is among British Columbia's best cottage wineries. Its Mediterranean-style facility, surrounded by pine woods and vineyards, is set on a hill overlooking Lake Okanagan. The winery is owned and operated by Guelph University–trained oenologist Bob Claremont, who worked for British Columbia wineries Mission Hill and Calona before buying thirty-five acres of well-established vineyards outside of Peachland.

Claremont's vineyards are under cultivation to viniferas and British Columbia hybrids such as Okanagan Riesling. From these grapes and additional crops purchased locally, he produces 20,000 gallons of table wines annually. Of these, Claremont is best known for Sauvignon Blanc (a gold medal winner) and Okanagan Riesling (a silver medal winner).

Tours of the winery are regularly conducted in the summer; they begin in the vineyard where picking and crushing are discussed. Next you will see the storage area with its carved oak barrels, the bottling room, and warehouse facility. As is customary, tours conclude in the tasting room, where Claremont's nine varieties of wine, including two reds, two rosés, and seven whites, can be sampled. You might enjoy a picnic lunch in the winery's lovely brick courtyard; if you don't bring your own food, supplies can be obtained in nearby Peachland.

Open all year: Monday-Saturday, 10-4:30. Guided tours: summer months, 11, 1, and 3; rest of year, by appointment. Free tasting (with or without tour). Retail outlet: wine, wine accessories, gifts. Arts and crafts show (Mother's Day weekend). Picnic facilities. Access for the handicapped. Directions: On Trepanier Bench Road, 1 mile off Highway 97 south, just north of Peachland.

Penticton

Casabello Wines
2210 Main Street
Penticton, British Columbia V2A 5H8
Telephone: (604) 492-0621
Owner: Ridout Limited

Casabello Wines is located in an attractive Mediterranean-style building on Penticton's Main Street. Built by former hotelier Evans Lougheed and a group of investors, Casabello is one of British Columbia's largest wineries, with sixty-six wines on its product list and a storage capacity of 3-million gallons. The winery has become a popular tourist attraction in the Okanagan Valley; its comprehensive guided tours last an hour and are extremely informative. In summer months, tours leave every twenty minutes; they begin at the winery building and then proceed to the fermentation room (this is only shown during the harvest), the processing room (the winery utilizes a centrifuge filtering system), and the bottling line. Afterwards you are shown through a one-acre demonstration vineyard and then shown a film which explains the art of winemaking. Three wines can be sampled in the tasting room; there is also a retail store on the premises with a good selection of wine-related accessories for sale. Casabello makes Canada's leading light wine, Capistro. It also produces varietal wines from Gewürztraminer, Johannisberg Riesling, Pinot Noir, and Chardonnay grapes.

There are no picnic facilities on the premises, but since this is an arid, semi-desert area with temperatures ranging from 90 to 115 degrees Farenheit in summer, after your visit you may want to take your hamper and head for the nearby beaches of Lake Okanagan or Lake Skaha for a refreshing dip and lunch.

Open all year. May-September: Monday-Saturday, 10-6; Sunday, 12-6. Other months: Tuesday-Saturday, 12-6. Guided tours: spring and fall, Monday-Friday, 11 and 2; July-Labor Day weekend, daily, every twenty minutes from 9:40 to 4. Free tasting. Retail outlet: wine and wine-related accessories. Directions: From Highway 97, take turnoff at Penticton sign (winery in town center).

Westbank

Mission Hill Vineyards
Mission Hill Road (P.O. Box 610)
Westbank, British Columbia V0H 2A0
Telephone: (604) 768-7611
Owners: Nick Clark and Anthony Von Mandl

Set on a knoll overlooking Okanagan Lake and the town of Westbank, Mission Hill Vineyards' Mediterranean-style winery commands beautiful views of the valley below. Mission Hill was founded in 1966 by fruit grower R. P. Walrod. He died before its completion, however, and the winery was subsequently purchased by Ben Gintner who was unsuccessful in his attempts to establish a thriving business, though he changed the name several times in his attempt to find a winning formula. Its current owners, Nick Clark and Anthony Von Mandl, purchased Mission Hill from Gintner in 1981.

Clark and Mandl have revived original owner Walrod's intention to produce premium wines. They own no vineyards, but have 320 acres under contract; the product line focuses on vinifera varietals, among them Mission Hill Private Reserve (a numbered series, only available at the winery), which includes Gewürztraminer, Chardonnay, Johannisberg Riesling, Chenin Blanc, Pinot Noir, and Cabernet Sauvignon. But Mission Hill's European-trained winemakers also produce red and white generics and more widely available varietals under the Mission Hill Vineyards label, as well as sparkling alcoholic ciders from apples and pears, the latter particularly refreshing in hot weather. And they are currently aging brandy for later distribution. Mission Hill is particularly known for its Private Reserve Johannisberg Riesling and Chardonnay.

Visitors who have toured the Napa Valley wineries will be struck by Mission Hill's resemblance to these California facilities. But the white stuccoed facade, bell tower, arches, and antique furnishings are no coincidence, since Clark and Mandl, with a proven track record as Canadian wine importers, have intentionally patterned Mission Hill after the prestigious and successful Robert Mondavi winery.

Tours of the winery are given all year round. Guides will show you through the whole working winery, and afterwards you can usually sample four wines. If you've brought a picnic lunch, you can enjoy it at one of the tables on Mission Hill's verdant lawn.

Open all year: summer, 10-8; winter, 10-4. Guided tours: on the hour in summer; weekends only in winter. Free tasting (with or without tours). Retail outlet: wine and wine-related accessories. Picnic facilities. Directions: From Kelowna, go south across Okanagan Floating Bridge. At second set of traffic lights after bridge, turn left on Boucheri Road. Follow Boucheri for 5 miles to Mission Hill Road; make a sharp right on Mission Hill (you will see winery at top of the hill).

NOVA SCOTIA

Grand Pré

Grand Pré Wines
Highway 1 (P.O. Box 18)
Grand Pré, Nova Scotia B0P 1M0
Telephone: (902) 542-4482
Owner: Roger L. Dial

Grand Pré Wines does not fit the profile of the other wineries discussed in this book; tours are by appointment (or "by chance") according to the owner, and, because of Nova Scotia's local restrictions, you cannot taste or even buy wine on the estate premises. But it has been included, nevertheless, because this maritime province has a climate on its southwestern side, moderated by the Bay of Fundy, that makes it a promising vinous region. Grand Pré's Californian owner and winemaker, Roger Dial, a professor of political science at Dalhousie University in Halifax, was the first to recognize this grape-growing potential and he produces some of the most interesting red wine in Canada at his tiny estate winery.

Grand Pré Wines occupies a two-story farmhouse in this town of historic dwellings first settled by the Acadians in the 1680s. Evangeline National Park, with its camping sites and beach, is nearby. The winery's lower level, dating to the early 1800s, contains the cellar and production facilities; upstairs, a new addition houses the retail area (grapes only), laboratory, and offices. On the premises are five of Dial's twenty-six acres of estate vineyards. (The remaining vines are on a mountain in Billtown.) Here, in Nova Scotia's unusually hospitable climate—the island enjoys a longer fall growing season than Canada's most important wine district, the Niagara Peninsula—Dial grows Chardonnay, Gewürztraminer, Pinot Noir, Kerner, Bacchus, Maréchal Foch, Baco Noir, Seyval Blanc, and the

unusual Russian *vitis amurnesis*. From the latter Russian cultivar, Dial produces his most complex and sophisticated wine, a vintage-dated varietal Cuvée d'Amur.

Dial's winery is a small, traditional operation, employing tested viticultural techniques, including wood aging. Purposely small (Dial doesn't want Grand Pré's production to grow above six to eight thousand cases), it is a classic prototype of a cottage winery, and a very interesting place for the serious oenophile to visit. Tours are personally conducted by Dial or one of his sons. But to taste the fruits of his labor, you will have to purchase Grand Pré's product at the nearest government-owned N.S.L.C. Specialty Store.

Open by appointment or by chance. Guided tours. No tasting or sales (both are illegal in Nova Scotia on winery premises). Directions: From Halifax, Highway 101 towards Annapolis Valley to exit 10. From exit, right turn onto Highway 1. Follow Highway 1 for 1 mile to Grand Pré Crossroads (its four corners are marked by a gas station, grocery, and motel). Go through intersection (staying on Highway 1) and take second driveway on right to the winery.

ONTARIO

Blenheim

Charal Winery and Vineyards
Highway 3 at Highway 40 (R.R. 1)
Blenheim, Ontario N0P 1A0
Telephone: (519) 676-8008
Owners: Allan and Charlotte Eastman

Charal Winery and Vineyards is one of the newest and most rapidly expanding winegrowing ventures in the Niagara Peninsula's fertile wine region. The winery was founded in 1975 by Charlotte and Allan Eastman and is located on the 300-acre family farm at Blenheim (called Porky's Corner by local residents).

The Allan Eastmans are the third generation in their family to farm this land and since 1968 have planted more and more of its acreage to wine grapes. Their vineyard now encompasses 100 acres and twenty-one grape varieties, including Concord, Agawam, Niagara, Fredonia, Seyval Blanc, Chardonnay, Riesling, Pinot Noir, Maréchal Foch, and de Chaunac. From

these estate-grown grapes and additional purchased fruit, the Eastmans currently produce 30,000 cases of wine annually. Table wines account for eighty percent of their output with the rest divided equally between sparkling and fruit wines. Charal's wines have already garnered twelve international awards; the Chardonnay, Leon Millot, and Seyval Blanc are particularly noteworthy.

Charal's very modern production facilities are incorporated in a simple cement building surrounded by the estate vineyards and orchards. Nearby is the family's original vegetable stand, burgeoning with seasonal fresh produce and fruit, homemade condiments and cheeses, and cider. Also on the farm are a newly opened bakery and the tasting/retail area for the winery.

Tours, which take about an hour, are given daily from May to September at 2 p.m. And if you visit during the crush, you may see the combination mechanical harvester/destemmer-crusher in operation. (Although this type of field equipment is the standard in California, it is unusual in Ontario.) Afterwards, you are welcome to walk through the estate vineyards and orchards and picnic at one of the tables next to the winery.

Open all year, except major holidays: Monday-Saturday, 10-6. Guided tours daily: May-September at 2. Free tasting. Retail outlets: wine, gifts, produce. Picnic facilities. Access for the handicapped. Directions: From Toronto, southwest on Highway 401 to Highway 40. Southwest on 40 to Blenheim. Winery at intersection of Highways 40 and 3 in Blenheim.

Harrow

Colio Wines of Canada
Walker Road (P.O. Box 372)
Harrow, Ontario N0R 1G0
Telephone: (519) 726-5317
President: Enzo DeLuca

Large (800,000 bottles) modern winery built by Italian investors in 1980. Located in Harrow—Canada's most southerly town—due south of Charal Winery and Vineyards. Young vineyards nearby in Essex County; ninety-nine percent of grapes purchased from local growers. Product list: varietal and generic table wines from vinifera and French hybrids include Johannisberg Riesling, Maréchal Foch, and Seyval Blanc; best noted for Bianco Secco, a white generic. Italian winemaker, Carlo Negri. Open all year: Monday-Saturday, 10-5. Guided tours in summer; other months by appoint-

ment (one day's notice suggested). Retail outlet: wine. Access for the handicapped. Directions: From Windsor, south on Walker Road for 30 miles to winery (in the town of Harrow).

Kingsville

Pelee Island Winery
455 Highway 18 East
Kingsville, Ontario N9Y 2K5
Telephone: (519) 733-5334
Owners: Wolf Von Teichman, Helmut Sieber, Walter Strehn, Karl Yaki, and John Zepf

Pelee Island's new winery is located on the southern tip of the Niagara Peninsula just east of Kingsville (not far from Colio Wines). Its vinifera vineyards, the largest in Canada planted solely to European grape varieties, thrive offshore in the temperate climate of Lake Erie's Pelee Island. (Pelee Island is near the international boundary which divides Lake Erie, not far from Ohio's wine islands.)

Pelee Island's vineyards have been nurtured by Austrian viticulturist and winemaker Walter Strehn, who revived the once successful Vin Villa Vineyard. One of Ontario's principal wine producers 100 years ago, Vin Villa met its demise during Prohibition, when many of its vines were destroyed and replaced with more profitable agricultural crops. In the 1970s, Strehn and his partners began replanting the island's acreage to Pinot Noir, Riesling, Scheurebe, Kerner, Gewürztraminer, Rieslaner, and Chardonnay. Currently, 120 acres are under cultivation. From this harvest and additional purchased grapes, the winery produces approximately 100,000 bottles of table wine annually. Pelee Island is particularly known for its Late Harvest Eiswein and award-winning Pinot Noir.

Pelee Island Winery's handsome new Kingsville facility is constructed of stucco with stone accents. The tour offered here will give you an excellent one-hour introduction to winemaking in Ontario. It includes an audiovisual presentation, a short lecture on the history of winemaking in the region, a tour of the working winery—including processing equipment, aging cellars, bottling line, and warehouse—and a tasting of seven of the firm's wines.

Day trips via ferry to Pelee Island's vineyards are planned for seasonal months. These will include a tour of the island's various sights (via bus), with stops at the old Vin Villa winery, a local pheasant farm, and the vineyards. There will be time for swimming on the island's sandy beaches; lunch and wine will be served at the vineyards.

Open all year: Monday-Saturday, 9-6; Sunday, 1-5. Guided tours and tasting ($3): Monday-Saturday at 12, 2, and 4; Sundays at 1, 2, 3, and 4. Retail outlet: wine, wine and cheese baskets, gifts. (No wine is sold on Sundays.) Wine Festival (September). Directions: From Windsor, take Highway 3 east to Kingsville Division Road. Kingsville Division Road to Highway 18. Winery is located 1 mile east of Kingsville (past railroad tracks at four corners).

Niagara Falls

T. G. Bright & Co.
4887 Dorchester Road (P.O. Box 510)
Niagara Falls, Ontario L2E 6V4
Telephone: (416) 357-2400
President: Edward S. Arnold

T. G. Bright & Co., informally known as Brights, is Canada's largest winery and vineyard operation, and is second in age only to Barnes Wines in St. Catharines. Brights' capacity is a staggering nine million gallons.

The winery was founded in Toronto in 1874 by two enterprising young men, T. G. Bright and F. A. Shirriff. Sixteen years later Bright and Shirriff moved their winery to Stamford Township (now called Niagara Falls) to be in Ontario's grape-growing region. (The winery is located just a ten-minute drive west of the Falls; during the late 19th century, it shared the same name as the popular tourist attraction, for it was called the Niagara Falls Wine Company.) In 1910 the Bright family bought out Shirriff's interests and the company received its current name.

Throughout its long history, Brights has been a pioneer in the Canadian wine industry, with a list of top awards that would easily fill a number of pages. But its greatest contribution has been its grape cultivation program, initiated under French-born winemaster and chemist Adhemar de Chaunac. De Chaunac's experiments with imported varieties of hybrids and viniferas produced no fewer than thirty successful hybrids. His contribution to viticulture was acknowledged when the de Chaunac grape was named for him.

As befits its stature, Brights has particularly inviting visitor facilities. Its hospitality center is constructed of two massive 68,000-gallon wooden tanks. Tours of the winery, located beyond, begin here; they generally take an hour, including a film and tasting. Much of the winery's original cooperage is still used to age the red wines; immense stainless-steel tanks hold the whites.

Brights has an incredibly long product list. More than seventy-five labels encompass almost every category from aperitifs, reds, and rosés to whites, sparkling wines, and dessert wines. The firm's wines have been honored with many awards. The Baco Noir has received five gold medals and President Canadian Champagne is extremely popular.

Open all year: Monday-Saturday, 9-5. Guided tours: May 1 - October 31 at 10:30, 2, and 3:30; November 1-April 30 at 2. Free tasting. Retail outlet: wine, gifts. Access for the handicapped. Directions: From downtown Niagara Falls, take Highway 420 to Dorchester Road. Follow Dorchester north to winery (on your left).

Niagara-on-the-Lake

Inniskillin Wines
Line 3 (R.R. 1)
Niagara-on-the-Lake, Ontario L0S 1J0
Telephone: (416) 468-2187
Owners: Donald J. P. Ziraldo and Karl J. Kaiser

Inniskillin, located in the northeast corner of Ontario's Niagara Peninsula, is the province's first estate winery to be bonded since 1929. Its name honors the Inniskillin Fusiliers, who were stationed in the Niagara area during the War of 1812 and are now part of the British Armed Forces. Inniskillin's vineyard, established in 1971 by nurseryman Donald Ziraldo, has since been expanded to thirteen acres on which are planted both vinifera and hybrid grape varieties. Its new winery building is on Brae Burn Farm, adjacent to the vineyard. Skillfully designed by architect Rapheale Belvedere, it blends with the site's picturesque weathered barns; one (dating to the early 1900s) has recently been handsomely restored to serve as Inniskillin's wine boutique.

Inniskillin produces 250,000 gallons of wine annually. The extensive product line, consisting of both table and *methode champenoise* sparkling wines, includes estate-bottled varietals (Chardonnay, Gamay, Seyval Blanc, Riesling, Maréchal Foch, and Chelois) and generics (Vin Nouveau and Brae Blanc). All wines are vintage-dated and many are vineyard-designated; the 1980 Maréchal Foch is particularly noteworthy.

While there are no formal tours of the winery, you are welcome to look around its production areas. These include the barrel-aging cellars, champagne loft, and outdoor crush pad. Usually six wines are available for sampling in the nearby wine boutique; numerous educational exhibits are mounted on the walls of its four rooms. These cover the history of the winery, cork production, the winemaking process, and other related topics.

Open all year: Monday-Saturday, 10-6. Self-guided tours. Free tasting. Retail outlet: wine, wine accessories, and locally made wine jelly. Directions: Queen Elizabeth Way to Highway 55. Highway 55 northeast to Line 3. East on Line 3 (past Creek Road) to winery.

Reif Winery
Niagara Parkway (R.R. 1)
Niagara-on-the-Lake, Ontario L0S 1J0
Telephone: (416) 468-7738
Owner: Ewald Reif

Newest (opened in 1983) and one of the smallest of Niagara-on-the-Lake's estate wineries. Retail sales, tasting, and production facilities in a renovated stagecoach building. Owner/winemaker Ewald Reif's German forebears began making wine more than a century ago. Reif's 130-acre vineyard planted to thirteen grape varieties, including Villard Noir, de Chaunac, Maréchal Foch, and Seyval Blanc. Product line includes seven 100-percent varietal wines and two blends, all produced from estate-grown grapes and aged in French oak. Particularly noted for Rieslings and Siegfried Rebe. Open all year: Monday-Saturday, 10-5. (Summer: 10-6.) Guided tours: weekdays at 10:30, 1:30, and 3:30; Saturdays at 10:30, and on the hour from 1 until closing. Free tasting (with or without tour). Retail outlet: wine. Grape and Wine Festival (last twelve days of September). Directions: From Niagara Falls, follow Niagara Parkway north towards Niagara-on-the-Lake. Winery on Parkway just past Line 3 exit.

St. Catharines

Barnes Wines
Martindale Road (P.O. Box 248)
St. Catharines, Ontario L2R 6S4
Telephone: (416) 682-6631
President: Jeffrey Ward

Barnes Wines, in operation since 1873, is Canada's oldest winery. Located northwest of Brights Wines, whose operation is nearly as venerable, the winery's complex of old and new buildings sits on the bank of the picturesque Welland Canal which originally linked Lake Ontario and Lake Erie. The winery was founded by George Barnes, who called it "the Ontario Grape Growing and Wine Manufacturing Company." His choice of a lakeside location not only was convenient to St. Catharines' many vineyards, but turned out to be a marketing plus as well. For as the sailing

CELLAR
RESERVE

1983
Seyval Blanc

WHITE WINE • VIN BLANC

BARNES WINES LIMITED, ST. CATHARINES, CANADA
PRODUCT OF CANADA • PRODUIT DU CANADA
750 mL 11.5% alc./vol.

vessels waited their turn to be pulled through the canal by mule barge, the ships' captains would often tie up at Barnes's wharf, taste his wine, and frequently buy a barrel to take on the long voyage ahead. By 1894, the winery could store 250,000 gallons, most of it kept in stone cellars cut twenty feet below ground level.

Today Barnes produces 270,000 cases of wine annually and has a storage capacity of 1.5 million gallons. Sixty percent of this is table wine; the other forty percent is divided equally between sparkling and fortified wine. Twenty acres under cultivation at the winery site are planted to Seyval Blanc, de Chaunac, Maréchal Foch, Johannisberg Riesling, Chardonnay, Elvira, and Concord, but Barnes purchases the bulk of the fruit—2,000 tons—needed for production.

Visitors to Barnes should start at the winery's original cut-stone building; its deck is constructed from the staves of old wine vats. Within are housed the visitor center, with its interesting museum of winery artifacts dating to the mid-19th century, the tasting room, and the retail store. Tours of the winery begin with an audiovisual presentation covering the Niagara grape and wine industry. Your visit to the cellars in the modern adjoining winery is accompanied by an explanation of winemaking techniques; the vineyard is visible through a number of windows. On your return to the tasting room, a structured sampling of five Barnes wines is conducted; commentary on wine appreciation and tasting is supplied by staff members.

The Barnes product list includes 100-percent vintage-dated varietals, blended wines, and four-liter "bag-in-box" wines. The winery is particularly known for its "Heritage Estates" wines; these include five blends— Chablis, Burgundy, Claret, Rose, and Rhine Wine—and four sherries. Both Johannisberg Riesling and "Adagio" have been gold-medal winners in European competitions.

Open all year for combined guided tour/free tasting. June-October: Monday-Saturday, 11, 1, and 3; November-May: Saturday, 1 and 3. Wine Museum. Niagara Grape and Wine Festival (last ten days in September). Access for the handicapped. Directions: From Niagara Falls, Queen Elizabeth Way to St. Catharines. Exit at Martindale Road and proceed south one block to winery.

St. Davids

Chateau Des Charmes Wines
Four Mile Creek Road (P.O. Box 280)
St. Davids, Ontario L0S 1P0
Telephone: (416) 262-4219
Owners: Paul M. Bosc, Rodger A. Gordon, Larry Needler

Second largest (125,000 gallon capacity) of the estate wineries in the Niagara-on-the-Lake area. Functional cement-block winery surrounded by 100-acre vineyard, established in 1978, planted to Chardonnay, Riesling, Gamay Beaujolais, Aligoté, Auxerrois, Pinot Noir, Villard Noir, Gewürztraminer, Cabernet Sauvignon, Merlot, Cabernet Franc, Seyval Blanc, and Vidal Blanc. Product list: estate bottled varietal and generic wines including Pinot Noir, Cour Blanc and Rouge, Sentinel Blanc and Rouge, Entre Nous White and Red. Dijon-trained French winemaker, Paul Bosc, best known for his vinifera varietals. (Winery named for former Bosc family's villa, Charmes, on the Algerian coast.) Open all year: Monday-Saturday, 9-5. Guided tours by appointment at 11, 1:30, and 3 (five days' notice requested). Free tasting (with or without tour); all wines available for sampling. Retail outlet: wine. Access for the handicapped. Directions: Queen Elizabeth Way to Highway 55 exit. Follow Highway 55 north to Old Highway 8. East on Highway 8 to St. Davids. North from St. Davids on Four Mile Creek Road to winery (on your right).

Winona

Andres Wines
South Service Road (P.O.Box 550)
Winona, Ontario L0R 2L0
Telephone: (416) 643-4131
President: J. A. Peller

Largest vintner in Canada with plants in six provinces. Winona winery in sprawling one-story modern building with barrels stacked out front. Just outside of Grimsby near Lake Ontario. Extensive list includes more than fifty varieties; Andres made its reputation on Baby Duck; also known for Hochtaler. Popular guided tour (1½ hours with tasting; reservation required) includes slide presentation, tour of the physical plant, and tasting (whole line can be sampled). Open for tour April-November: Monday-Saturday, 9-9. Open all year for tastings and sales. Retail outlet: wine, cider, gift packs. Public conservation park across the way for picnicking. Limited access for the handicapped. Directions: Queen Elizabeth Way north towards Grimsby. Exit at Casa Blanca Blvd. Take South Service Road to winery.

 # Suggested Reading

General Resources

Adams, Leon D. *The Wines of America*. 3rd edition. New York: McGraw-Hill Book Company, 1985.

Aspler, Tony. *Vintage Canada*. Scarborough, Ontario: Prentice-Hall Canada Inc., 1983.

Blumberg, Robert S. and Hurst Hannum. *The Fine Wines of California*. 3rd edition. Garden City, New York: Doubleday & Company, Inc., 1984.

Cattell, Hudson and Lee Miller. *Wine East of the Rockies*. Lancaster, Pennsylvania: L & H Photojournalism., 1982.

Fegan, Patrick W. *Vineyards & Wineries of America*. Brattleboro, Vermont: The Stephen Greene Press, 1982.

Johnson, Hugh. *Hugh Johnson's Modern Encyclopedia of Wine*. New York: Simon and Schuster, 1983.

Moore, Bernard. *Wines of North America*. London: Winchmore Publishing Services Limited, 1983.

Robards, Terry. *Terry Robards' New Book of Wine*. New York: G. P. Putnam's Sons, 1984.

Tartt, Gene, Ed. *The Vineyard Almanac & Wines Gazetteer*. Saratoga, California, 1984.

Wagenvoord, James. *The Doubleday Wine Companion, 1983*. Garden City, New York: Doubleday & Company, Inc., 1983.

Regional Travel Guides and Pamphlets

California's Wine Wonderland. San Francisco: Wine Institute, 1983.

Church, Ruth Ellen. *Wines of the Midwest*. Athens, Ohio: Swallow Press Books (Ohio University Press), 1982.

Discover Oregon Wines & Wineries. Portland, Oregon: Oregon Winegrowers Association, 1984.

Giordano, Frank. *Texas Wines & Wineries*. Austin, Texas: Texas Monthly Press, Inc., 1984.

Holden, Glenda and Ronald. *Touring the Wine Country of Oregon*. Seattle, Washington: Holden Travel Research, 1982.

_____. *Touring the Wine Country of Washington*. Seattle, Washington: Holden Pacific, 1983.

Taste the Wines of Pennsylvania and . . . Taste Tradition in the Making. New Hope, Pennsylvania: Pennsylvania Wine Association, n.d.

Thompson, Bob. *Guide to California's Wine Country.* 3rd edition. Menlo Park, California: Sunset Books (Lane Publishing Co.), 1982.

Virginia's Wine Country. Richmond, Virginia: Virginia Division of Tourism, 1984.

Yamhill County Wineries. McMinnville, Oregon: McMinnville Chamber of Commerce, n.d.

List of Wineries and Vineyards

A

Alba Vineyard, 59-60
Alexander Valley Vineyards, 182-83
Almaden Vineyards, 212-13
Anderson Valley Vineyards, 137-38
Andres Wines, 263

B

Alexis Bailly Vineyard, 109
Barboursville Winery, 86-87
Bargetto Winery, 216-17
Barnes Wines, 260-62
Beaulieu Vineyard, 162-63
Benmarl Vineyards, 29
Beringer Vineyards, 166-67
The Biltmore Estate Wine Company, 81-83
Binns Vineyard and Winery, 139
Boordy Vineyards, 53-55
Boskydel Vineyard, 105
Bridgehampton Winery, 47-48
T.G. Bright & Co., 257-58
Brights House of Wine, 248
Brotherhood Winery, 31-32
Bucks Country Vineyards & Winery, 69-70
Buena Vista Winery, 192-93
Bully Hill Vineyards, 39-40

C

Cakebread Cellars, 163
Callaway Vineyard & Winery, 222-23
Canandaigua Wine Company, 35
Casabello Wines, 250
Casa Larga Vineyards, 38-39

Cascade Mountain Vineyards, 25-27
Catoctin Vineyards, 53
Cedar Hill Wine Company, 119
Chaddsford Winery, 65-66
Chalet Debonné Vineyard, 121
Charal Winery and Vineyards, 254-55
Chateau Benoit Winery, 227
Chateau Des Charmes Wines, 262
Chateau Esperanza Winery, 33-34
Chateau Grand Traverse, 108
Chateau Montelena Winery, 150-52
Chateau St. Jean Vineyards and Winery, 190-91
Chateau Ste. Michelle, 239-41
Chicama Vineyards, 20-21
Christina Wine Cellars, 131-32
Claremont Estate Winery, 249
Clos Du Val, 156-57
Colio Wines of Canada, 255-56
Columbia Winery, 234-35
Commonwealth Winery, 18-19
Concannon Vineyard, 207-208
Congress Springs Vineyards, 215-16
Creston Manor Vineyards and Winery, 218
Crosswoods Vineyards, Inc., 15
Cuvaison Vineyard, 152-53

D

Bernard D'arcy Wine Cellars, 56-57
Dehlinger Winery, 191
Domaine Chandon, 174-76
Door Peninsula Winery, 134
Duplin Wine Cellars, 83-84

267